The Wisdom of Creation

Barbara E. Bowe, R.S.C.J.
Mary C. Boys, S.N.J.M.
Walter Brueggemann
Agnes Cunningham, S.S.C.M.
Carol J. Dempsey, O.P.
Mary Frohlich, R.S.C.J.
Anthony J. Gittins, C.S.SP.
Mary Catherine Hilkert, O.P.
Andrew L. Nelson
Hermann E. Schaalman
Richard J. Sklba

Edited by
Edward Foley, CAPUCHIN
Robert Schreiter, C.PP.S.

A Michael Glazier Book

LITURGICAL PRESS
Collegeville, Minnesota

www.litpress.org

A Michael Glazier Book published by the Liturgical Press

Cover design by David Manahan, O.S.B.

1	2	3	4	5	6	7	8

Library of Congress Cataloging-in-Publication Data

The wisdom of creation / Barbara E. Bowe . . . [et al.] ; edited by Edward
   Foley and Robert Schreiter.
       p. cm.
    "A Michael Glazier book."
    Includes bibliographical references and index.
    ISBN 0-8146-5122-4 (pbk. : alk. paper)
    1. Creation. 2. Wisdom literature—Criticism, interpretation, etc.
3. Ecology—Biblical teaching. 4. God—Wisdom. I. Bowe, Barbara
Ellen. II. Foley, Edward. III. Schreiter, Robert J.

BT150.W57 2004
231.7'65—dc22

                                    2003022673

To
Dianne Bergant, C.S.A., PH.D.,

professor of biblical studies,
Catholic Theological Union, Chicago,

biblical guide,
friend of creation,
woman of wisdom.

Contents

Introduction

There has been a growing awareness of the plight of the earth at the hands of human beings over the past several decades. Starting with Rachel Carson's *Silent Spring* (1962), an increasingly insistent chorus of voices has been calling the human family to a greater attentiveness and care for the earth and its resources, in the face of pollution, the extinction of animal and plant species, and general environmental degradation. In the 1970s, the World Council of Churches took up this summons in its program "Justice, Peace, and the Integrity of Creation." The Roman Catholic Church, under Pope John Paul II, has also joined in the efforts to create awareness of the scale of the problem, as well as urge that steps be taken to slow down this erosion of the very basis for life on this planet.

The theological community is also making a response. The first voices to be heard were those which alerted people both to this misuse of God's creation and to the fact that the Scriptures had been used to justify a reckless human domination of the earth. What has become clearer in the course of time is that the need to reexamine the theological resources used to bolster such misuse had to go much deeper: it touches our understanding of ourselves as creatures of God, our understanding of the human being itself, and the way we do theology at all. The ecological crisis calls for a shift in the way theology is undertaken, i.e., how it reads its sources, what counts for a defining argument, and a close scrutiny of the consequences of theological positions and statements.

An important contribution to this discussion over the past two decades has been the work of Dianne Bergant, C.S.A. She has woven together two strands of thinking that now provide part of the warp and woof of a new way of doing theology in the service of care for the earth.

She brings first of all a deep concern for the earth and its creatures. In a series of articles and books, this theme has echoed through her writings. As a professor of biblical studies at Chicago's Catholic Theological Union, she has brought a second resource to this discussion: the contribution of the Wisdom literature of the Jewish tradition. This body of writings, coming late in the First Testament period, carries with it a concern for right relationships, for attention to the consequences of one's actions, and a contemplative and reverent stance toward God's creation. It is at the intersection of these two resources—concern for the earth and the Wisdom literature—that much of her work has been done.

This volume of essays, written by colleagues and friends to honor her twenty-five years on the faculty at Catholic Theological Union, takes up the themes of Creation and Wisdom from a variety of different perspectives, both biblical and theological, to think along with Bergant about the challenge of care for the earth and those who dwell upon it.

The first six essays look at different dimensions of wisdom in the Hebrew and Christian Scriptures. In the opening essay Walter Brueggemann explores the kind of knowing which wisdom brings to our understanding of creation. Wisdom, he notes, has a theology of creation as its framework. He presents a novel approach to what wisdom brings to our understanding of creation by focusing on two texts: Isaiah 1:2 and Jeremiah 8:7. Both of these texts point to an innate ability of creatures to relate to their Creator. Specifically, he notes what the nonhuman creatures know, and what the human creatures are slow to learn. With help from an imaginative novel and a whimsical essay, he asserts that the knowledge nonhuman creatures have of their Creator is not somehow primitive, but rather a "responsive creatureliness that adheres to the uncompromising reality of the Creator." This provides not just an interesting way to read the Bible, but constitutes a "resilient alternative" to a way of thinking that sees humans as exceptional from the rest of the creation. Rather, the way of wisdom thoroughly embeds human beings in the midst of creation. Humans have a lot of learn from the rest of creation—and are perennially slow at doing so.

Richard J. Sklba continues this line of thought by an examination of the book of Proverbs. He notes that wisdom literature often arises at moments of epochal change in societies. Such wrenching change was going on in Israel in the two centuries before the birth of Christ. What Jesus ben Sirach did in his collection of aphorisms and sayings was to distill for Israel the wisdom of its past, and the new insights emerging in the spreading Hellenism of the period. Sklba focuses upon the image of the table and the feast as the locus for transmitting wisdom and seeing the effects of wisdom—as well as its negative counterpart, folly. Wisdom brings with it transformation and conversion, a new way of thinking and acting. It allows one to "think outside the box" and come to a different way of living in the world. Wisdom is about more than the refinements of table etiquette. It shows us how to live, how to relate, and how to lead. Certainly this message from wisdom's table is critical for a world that needs to change its perspectives and its way of living.

Part of that change in perspective and living requires a redefinition of the self, both in relationship to one's own self and to others. This has been a major agenda item of feminism in theology over the past three decades, and Dianne Bergant's writings have been a significant contribution in this regard. Mary C. Boys, s.n.j.m., takes up a topic that has become prominent in recent feminist readings of the Scriptures, namely, the recurring tendency in some feminist writing to read the story of Jesus from an anti-Judaic perspective. She calls this a theological "virus" that works often inadvertently to skew readings of the Jesus story, and thus creates a "catechesis of vilification" to define the Christian self over against Judaism. Such anti-Judaic readings, of course, go far beyond feminism. Boys is specifically concerned that seeking liberation for women must not be done at the cost of other (oppressed) groups. The defining of Jesus over against Judaism creates a false binary since Jesus himself was a Jew.

The exploration of this theme is important for the wider argument of this book from two perspectives. It is first of all about overcoming the vilification of Jews by Christians, a problem that has caused much suffering for the Jewish community over the last two millennia. But it is also about how we define ourselves. Humans have often defined themselves over against creation, to the detriment of the rest of creation. How we see the "self" and the "other" is key to the kind of redefinition that has to go on in theology for the sake of the preservation of the earth. Again, these two perspectives have been prominent in Bergant's work. For many years she has been active in the Catholic-Jewish dialogue and clearly concerned with these issues of definition.

Barbara E. Bowe, r.s.c.j., looks at two wisdom themes in the Gospel of John: "Woman Wisdom" and the "I am" sayings of Jesus. Bowe notes that the wisdom tradition "finds God revealed at the very heart of the world, in the ordinary daily common sense of proverbial wisdom, and in the midst of the great mysteries of life, such as suffering and death that touch all peoples everywhere." As such, it offers a broader, more "cosmopolitan" and "ecumenical" view of the world than a focus on human history alone. This has become an important contribution of the study of the wisdom literature to intercultural and interreligious discussions, since many traditions have this cultivation of wisdom.

Bowe traces how the wisdom tradition, presented as "Woman Wisdom" in the book of the Wisdom of Solomon, and in the "I AM" naming of God, help shape an understanding of Jesus in the j. She explores a host of wisdom motifs in that Gospel that retrace the story of Woman Wisdom as found in the book of the Wisdom of Solomon. She also examines Jesus' ascribing "I AM" to himself in a number of Johannine passages as well as the many "image" sayings found throughout the Gospel. The themes of wisdom in the Gospels have received renewed attention in recent years and, for a renewed theology, provide an important link between the larger wisdom tradition in the Scriptures and the Jesus story.

Herman E. Schaalman revisits the opening chapters of the book of Genesis, and considers how they reveal the complex and intriguing relationship between God and human beings. He begins by exploring the fact that Genesis does not record the creation of *Shabbat* or the day of rest until Adam has been created. He asserts that God does not appear here as self-sufficient, but "inherently, unavoidably" linked to us. This theme is developed further by exploring the deep involvement of humanity in creation itself (especially in the second creation account). Both of these reflections develop further a non-binary definition of the self—be it the human self in terms of God or in terms of the rest of creation. Schaalman's final statement captures this point succinctly: "The fundamental verity of being human is that it is a collaborative event between God and us." One can only understand human beings in terms of relationship.

The essay by Carol J. Dempsey, o.p., concludes these reflections on wisdom and creation in the Bible by giving us an overview of these themes running throughout the Scriptures. By doing so she models a conversation between the Bible and theology which will be extended through the remaining essays in this book. She looks not only at some of the familiar themes related to ecology—e.g., interdependence and care-giving instead of exploitation—but also at how creation relates to revelation and to redemption: themes which have received far greater attention in theology than have creation and wisdom. By doing so, she hopes to point toward a new paradigm of creation as a horizon for biblical theology. Many of those writing on religion and ecology have noted the need for such a paradigm, and Dempsey's essay helps make an important move in that direction.

Moving from consideration of the Scriptures to an exploration of the Christian spiritual tradition, Agnes Cunningham, s.s.c.m., examines three Christian figures noted for their love of wisdom: Clement of Alexandria, St. Augustine, and St. Louis de Montfort. She notes how each, in different contexts, reflect a search for wisdom that weds the wisdom of God with the wisdom to be discovered in creation. Clement, sometimes called the first Christian humanist, explored Christian wisdom, especially as found in the figure of Jesus Christ, in the rich and intricate world of *Gnosis* in Alexandria, the intellectual center of the world at that time. St. Augustine of Hippo wrote much about wisdom, notably in the *Confessions,* where we follow the quest for knowledge in his youth. Cunningham presents an Augustinian synthesis of wisdom in three strands seen running through Augustine's work: natural wisdom, theological wisdom, and mystical wisdom.

In St. Louis de Montfort, we have a figure who rediscovered wisdom in the midst of the post-Tridentine reform in France. This was expressed graphically in his design of the Wisdom Cross of Poitiers, meant to remind those who saw it of the wisdom of the cross, and the Eternal and Incarnate Wisdom who hung upon it. De Montfort's passionate search for wisdom in the form of the cross of Christ reveals to us yet another side of wisdom, i.e., the wisdom of God that is contrary to the wisdom of the world (cf. 1 Cor 1:23-25).

Anthony J. Gittins, c.s.sp., extends the spirituality dialogue by moving outside Christianity and addressing a question much in discussion today, namely, the revelation of God in other religions. As has already been noted, the wisdom traditions of different religions have become an important meeting point for interreligious dialogue today. Gittins takes us on an exploration of the "other faces of God" to be found in traditional religions in Africa in the hopes of finding out what these traditions can reveal about the nature of God. In the process, he reminds us that geography carries theology as well. The contours of the landscape shape our intellectual horizons as much as our epistemologies do. Echoing Augustine, he presents the world as "a theology book, or at least an illustrated essay about the Creator." Creation is an important communication act of the Creator, and it behooves us to read it in order to deepen our knowledge of God.

Mary Frohlich, r.s.c.j., in looking at the spirituality which will sustain people in the postmodern world, takes up the image of the garden as it is found in the Scriptures: specifically, the Garden of Eden, the lovers' garden in the Song of Songs, and the Garden of Olives in the Gospels. Gardens represent places of cultivation where the human being and the rest of creation come together. Our experience of gardens can help us relocate ourselves in the fragmentation of the postmodern world, where globalization has dislocated us, and pluralism has disoriented us. Frohlich leads us through reflections on these garden stories in order to chart a path of spiritual ministry for people living in a postmodern world.

Theology is not simply about reflection upon experience and tradition; it is also about communication. Good theology should translate into effective preaching. This too has been a motif of Dianne Bergant's work. She has authored a three-volume commentary on the Sunday lectionary readings and, since 2002, has provided the weekly commentary on the Sunday readings for the Jesuit weekly magazine *America*. The final two essays in this collection take up the challenge of preaching. Mary Catherine Hilkert, o.p., explores what it means to preach from the Book of Nature. She explores Nature as revelatory of God's word for us, recalling again Augustine's concern for finding the traces of God *(vestigia Dei)* throughout creation. She then takes up the groanings of nature under the oppression of injustice and exploitation, noting Leonardo Boff's injunction that the exploitation of human beings regularly leads to the plundering of nature as well. To preach from the Book of Nature entails preaching a "new logic of grace," in which right relationships are maintained and the glory of God is revealed in the extravagance of creation. The future of the earth requires, therefore, that we preach from what God has revealed in Nature as well as what God reveals in human history.

Andrew L. Nelson expands on the nature and significance of preaching. He examines especially the danger of moralizing rather than helping people move toward moral responsibility. Critics of recent ecological thought have often mentioned that writing in this area sometimes amounts to no more than a jeremiad intended to heighten a sense of guilt, rather than providing constructive ways of addressing issues. Nelson helps us find the way beyond such moralizing toward such constructive engagement.

Taken together, these essays help move us forward in creating a paradigm for responsible theology in the face of the threats to the future of the earth. They open up the themes of creation and wisdom from biblical, spiritual, and praxis perspectives which have too often been neglected in favor of redemption, history, and prophecy. They stress that how we choose to define ourselves in relationship to creation will be more revelatory than any categorical statements we try to make. The importance of images—of God, of ourselves, of creation, of the garden—communicate these concerns sometimes more effectively than discursive prose. And how we communicate will also be reflective of what we really believe.

The work of a theology that will contribute to the care of and the future of the earth is, in many ways, still in its early stages. Dianne Bergant's own work on themes of wisdom, creation, prayer, and preaching has made a significant contribution to that theology. It is hoped that these essays written in her honor will continue that important task of envisioning a better world, a vision we hope will grow and mature through future generations.

The Creatures Know!

WALTER BRUEGGEMANN

It is by now a truism that "wisdom thinks resolutely within the framework of a theology of creation."[1] That now common assumption among interpreters, however, has not always been obvious. It is, rather, a hard-won consensus that emerged in a season of scholarship preoccupied with "history," in which theological interpretation of the Old Testament was dominated by the programmatic slogan, "God acts in history." The connection between wisdom and creation has permitted interpretation to move outside "history" and to challenge the fear of "natural theology" that pertained in Barthian circles of interpretation. Once that consensus judgment was reached about creation and wisdom, however, it has not been at all clear what the creation-wisdom convergence may mean, and it has been treated very much in a reified way, either as a dogmatic claim, or as a method, or as a defense of "natural theology" as distinct from and opposed to a "theology of the word."

In this brief discussion I wish to explore the creation-wisdom convergence in a much more elemental way by considering that God's "creatures," the counterpoint to the "creator"—all of them, nonhuman as well as human—have a wisdom about how to live well, faithfully,

---

[1] Walther Zimmerli, "The Place and Limit of the Wisdom in the Framework of the Old Testament Theology," *Studies in Ancient Israelite Wisdom,* ed. by James L. Crenshaw (New York: KTAV Publishing House, 1976) 316.

and responsively in a world governed by the creator God. From the outset of my formulation of this brief argument, I have intended to consider two texts in particular that evidence the *knowing* of which nonhuman creatures are capable, Isaiah 1:2 and Jeremiah 8:7. While I have pondered this writing assignment, I was clear about the argument I wished to present to honor Professor Bergant; I have, however, puzzled for a long time about how to proceed. I finally found my way with the suggestions of a playful novel and a whimsical essay that relate loosely to our theme. It occurred to me that because sapiential texts are characteristically "figured" and not straight-forward, the two contemporary texts to which I will appeal, marked by whimsy and playfulness, are appropriate access points for my argument. I imagine that the ancient wisdom teachers who reflected upon the inscrutability of creation might approve such a slightly tongue-in-cheek approach that articulates created reality alternatively, an alternative that nonetheless lives close to the creaturely character of the world.

I.

The playful novel to which I refer and by which I have been given an opening is *Ishmael* by Daniel Quinn.[2] I first learned of the book recently from an e-mail from a west coast conservationist, whom I do not know, who wrote to ask me about the book. My correspondent reported that the book urges that the narrative of Genesis 4 is the clue to all contemporary social-political, environmental problems. By prompt return, without having seen the book, I expressed my misgiving about that thesis, given that the narrative of Genesis 4 functions in no important way in the remainder of the Hebrew Bible.

But then I read the book. The story line is that a gorilla, previously named Goliath but now renamed Ishmael, who becomes an "instructor" to a human student who has great curiosity. At first the two communicate by eye contact; but soon the gorilla speaks and leads his human student by sustained Socratic method to a revision of his notion of created reality.

It is the initial assumption of the human student, informed as Ishmael observes, by the story of "Mother Culture," that human beings stand outside the "laws of creation" and are capable and permitted to exercise

---

[2] Daniel Quinn, *Ishmael: An Adventure of the Mind and Spirit* (New York: Bantam Books, 1992).

control over the rest of creation as the exceptional creature with remarkable and absolute entitlements. Indeed, as Ishmael observes, the humanly constructed account of creation (as in Genesis 1–2) is to portray humanity as the ultimate and final creation, after which there will be none other. (It is observed in passing that the creation narrative as told by jellyfish make the same interpretive maneuver, that the goal and final act of creation is the emergence of jellyfish.)

Over time Ishmael begins to expose the mistaken narrative of human exceptionalism, and to present a critique of the notion that human persons stand outside the laws of creation as the final creature, entitled to control all other creatures. Ishmael identifies the human assumption of exceptionalism as the story of "Takers," those who believe that the proper human enterprise is to take land and food from all other creatures, to take more than one needs, to take it all, even at the expense of others; against that, Ishmael observes that in the long history of creation before the human Takers become dominant, that all creatures have been essentially "Leavers," that is, they take what they need of food, but not more than they need and, consequently, leave all the rest so that it may be used by other creatures who must also eat to live. Thus the long story told and practiced by all pre-human creatures is a story of leaving, a practice of leaving and life, a leaving of creation that makes sustained, sustainable life possible for the entire creation. Only quite belatedly in the long history of creation has the narrative of the Takers become dominant. That narrative, as it is vigorously embraced and practiced, is leading, in very short order, to the destruction of the viability of creation.

Ishmael articulates four things that the Takers do that jeopardizes all of creation, actions that would never occur elsewhere among non-human creatures:[3]

- The Takers exterminate their competitors;
- The Takers systematically destroy their competitors' food to make room for their own;
- The Takers deny their competitors access to food [In the wild the rule is: you may deny your competitors access to what you're eating, but you may not deny them access to food in general];
- Takers kill a surplus of food, but in the wild, animals never kill more than they can eat.

---

[3] Ibid., 126.

Ishmael formulates the "peace-keeping law" of the wild:

> You may compete to the full extent of your capabilities, but you may not hunt down your competitors or destroy their food or deny them access to food. In other words, you may compete but you may not wage war.[4]

Before he has finished, Ishmael affirms that Leavers have been content, for a very long time, to live "in the hands of the gods" who adequately provide food, whereas human creatures have come to know "good and evil," and so seek to secure their lives beyond reliance upon the gods:

> The premise of the Taker story is *the world belongs to man.* . . . The premise of the Leaver story is *man belongs to the world.*[5]

The Leavers make it possible for creation to go on forever, whereas the Takers pursue a policy that will bring creation to an end.

Before Quinn ends his book, he proposes, via Ishmael, that the agricultural revolution from hunting to planting, reflected in the narrative of Genesis 4, is the story of the Takers who now possess and own and produce their own food and thereby live against the law of peacemaking that makes life possible for all creatures.

The novel is, to be sure, lightly didactic. But Quinn resists decoding his story. It is clear that the Takers, the human community of exploitation and acquisitiveness, lives life against the fabric of nature and against the will of the creator. Ishmael's call for repentance, whether of Western humanity or of Enlightenment ideology, is repentance of the ideology of human exceptionalism, the recognition that humanity is subject to the same "peace-keeping laws" of the creator as are all other creatures, outside of which no sustainable life is possible.

It strikes me as peculiarly telling that Quinn has the truth of his story uttered to the human interlocutor by Ishmael, a gorilla. This comic narrative strategy asserts that Ishmael the gorilla, a hermeneutist for the nonhuman world in general, knows what his human partner learns only slowly, reluctantly, and with great obtuseness. Not too much should be made of the strategic way in which the novel works, except it cannot go unnoticed that the nonhuman agent in the narrative is the one who has wisdom about how the world works as an ordered, food-producing system that sustains life. The nonhuman agent knows what his human counterpart does not know and does not much want to

---

[4] Ibid., 129.
[5] Ibid., 239.

learn. We could as well take the next step for Ishmael to observe *that Leavers are wise in the ways of creation* and *that Takers practice a foolishness that inevitably leads to death.* Ishmael does everything for our subject short of using the word pair "wisdom-foolishness." Ishmael's advocacy is nicely lined out by his student's belated rejection of a familiar voice concerning the law of creation. The human partner to the learning process declares:

> Far and away the most futile admonition Christ ever offered was when he said, "Have no care for tomorrow. Don't worry about whether you're going to have something to eat. Look at the birds of the air. They neither sow nor reap nor gather into barns, but God takes perfect care of them. Don't you think he'll do the same for you?" In our culture the overwhelming answer to that question is "Hell no!" Even the most dedicated monastics saw to their sowing and reaping and gathering into barns.[6]

The speaker yearns to be out of "the hands of the gods," a yearning that is for Ishmael the ultimate foolishness.

II.

The whimsical essay by which I am informed is, "Creationism and the Spirit of Nature" by Peter Gabel.[7] This paper plunges directly into the "evolution-creation" debate, though with a particular and peculiar angle of vision. The author gives primary attention to the claims, and therefore the limitations of "scientific method." Such method, he observes, proceeds by a stance of "detachment and objectification" to such "objects" in nature through a scientific range of what is knowable. Gable suggests that such scientific method can only study the observable object and so stays on the "outside" of the object. Such method in principle cannot know anything from the "inside," and so misses out completely on whatever there might be of desire, will, or sense that can be known only intuitively and empathetically, from one knower to another knower. Thus the paper is an insistence that "scientific method" precludes knowledge about anything important that we should like to know about the "true existence" of a "living thing."

Gabel's case in point to which he refers repeatedly is the movement of a household plant toward the sunlight. Scientifically this movement

---

[6] Ibid., 228.

[7] Peter Gabel, "Creationism and the Spirit of Nature," idem, *The Bank Teller and Other Essays on the Politics of Meaning* (San Francisco: Acada Books, 2000) 45–67.

is to be understood by way of a theory of photosynthesis. Against such an "external" explanation, Gabel observes that the plant's movements are "unified," a sensual unity so that one "senses in the plant, the sense of pleasure that seems to manifest in the bend of the upper stem and the stretch of the highest leaves and that seems to contrast so strikingly with the droop of plants denied access to the same sunlight."[8] The author reports that his own sense "that the meaning of the plant's living movement exceeds the photosynthesis explanation is a very strong one."[9]

One must, so Gabel contends, step outside the detachment and objectification of scientific method in order to embrace "the plant in an intuitive movement of comprehension from one living being to another."[10] Gabel will not go so far as to say that the plant has "consciousness" or "will," but does insist that the plant is to be seen as "beautiful and good"—as miraculously alive and "here," no less than are we.[11]

In his critique of scientific method, Gabel has no patience with or appreciation for conventional "creationism." He makes clear his disdain for conventional creationism that is allied to unworthy right-wing causes. But he allows that

> creationists have been able to touch that dimension of people's ordinary experience that sense life in all its forms as expressive of some indwelling and miraculous beauty and goodness, and that knows with a certain intuition that this indwelling presence must be at the heart of any true knowledge of the world. However absurd the strict content of their views may be, and however evil may be the association of these views with right-wing militarism and anti-communism and with a servile dependency on fundamentalist preachers who purport to speak for an authoritarian God, there is something correct and admirable in their refusal to accept the hegemony of science as a privileged source of truth.[12]

In the end, Gabel's target is not creation and evolution as such, but the way in which decisions in politics and ethics are formed among us:

> The implications of what I am saying here go much deeper than the debate between evolution and creationism because if we could succeed in freeing knowledge from the grip of science and affirm the objectivity of

---

[8] Ibid., 46.
[9] Ibid.
[10] Ibid., 48.
[11] Ibid.
[12] Ibid., 52.

intuitive comprehension as the only route to understanding and com- municating about the *being* of things, we could also begin to transform the way people think about politics and ethics, about the meaning of their own lives and the lives of others and about what kind of world we should be trying to create.[13]

Real knowing in the end is intuitive, with respect as one "living thing" relates to other "living things." It is evident that Gabel, in his rather odd formulation, is concerned with "real knowledge" that concerns the in- scrutable character of "living things." Consequently he seeks to fend off foolish knowledge that imagines itself neutral and detached, foolish be- cause such a stance, so prized in our culture, precludes real knowledge.

III.

It is easy enough to entertain dismissal of these intellectual forays, because they lie outside established reason in our culture. Indeed, Gabel reports on such dismissive responses at the end of his article, though the responses also include some interesting affirmations. In the cases of both Quinn and Gabel, the writers reflect on the depth of creation that lies beyond scientific control. In Quinn's case, his attempt to understand creation also issues in an economic interpretation of the past in terms of Leavers and Takers. From the perspective of establish- ment rationality, articulation of "Leaver-Taker" on the lips of a gorilla is nonsense; and surely from that same perspective, Gabel's notion of empathetic, intuitive engagement is equally nonsense.

But, of course, in both cases the argument aims to subvert the es- tablished assumption that human persons with their much knowledge are free to act upon and, therefore, use all other creatures. In different ways, both Quinn and Gabel propose that pre-human *creatures know,* and what they know is congruent with the ordering of creation for life, an ordering that continues to elude human mastery. Both Gabel and Quinn ask the reader to step outside modern epistemological assump- tions and to entertain the thought—imagine!!—that nonhuman crea- tures know, perhaps know differently, perhaps know better, but in any case know the mystery of how the world works as God's generous, food-producing creation.

As I make a transition to consider biblical texts, I note in passing that the epistemology championed differently by Quinn and Gabel is not

---

[13] Ibid., 57.

unlike what was previously termed "primitive," "savage," or "dynamism" by historians of religion who considered the odd epistemological assumptions that seem to be reflected in the Bible. It is worth noticing, of course, that appeal to the work of Lucien Levy-Bruhl was taken to be a study in the evolution of thought toward rationality. The inescapable tone of those analyses was that this way of knowing is inadequate and inferior and it is, thankfully, only a step along the way to a more superior rationality of the modern mind.[14] It may be that we are still facing that contrast, but now we are perhaps sobered enough to entertain the thought that modern rationality cannot be so innocently championed, given the fact that it is, in Quinn's parlance, the work of "Takers," and in Gabel's categories the work of "detachment and objectification" that can never really know the inside of any living thing. It is evident that the Bible entertains another way of knowing that is not primitive or savage, but simply congruent with the character of the creation as the world over which God governs.

IV.

The first text that I will consider is Isaiah 1:3:

> The ox knows its owner,
>   and the donkey its master's crib;
> but Israel does not know,
>   my people do not understand.

This verse is preceded by an announcement of a lawsuit in which witnesses are called to observe the rebellious character of Yhwh's "sons" (v. 2). In our verse, the rebelliousness of these "sons," that is, Israel, is contrasted with the ox and the donkey who are not rebellious. The point of contrast is that for both subjects of the contrast the governing verb is "know." The contrast is simple and complete:

> Ox and ass *know,* Israel does *not know* (or discern).

What the animals know is the identity of their "owner" *(baal)* and the place of their reliable food. One can see this pragmatic knowledge by

---

[14] See William Foxwell Albright, *From Stone Age to Christianity: Monotheism and the Historical Process* (Baltimore: Johns Hopkins University Press, 1957) 168–78; also, Herbert F. Hahn, *Old Testament in Modern Research: A Comprehensive Synthesis of Modern Trends in Bible Studies* (Philadelphia: Muhlenberg Press, 1954) 59–74, 213–24.

observing such domesticated animals: at feeding time, toward dusk, they head home. They know what time it is and they know where their food is reliably provided. They know to whom they belong and on whom they can rely for food. It is likely that the usage of *"baal"* intends both to allude to the abundance of creation, and to set up a remarkable contrast between *baal,* a generic, non-covenantal food provider and the One to be known by Israel, the covenant Lord who presides over the food supply of the Torah community.

The appeal to such creaturely knowledge, of course, serves to accent Israel's lack of knowledge. If the verb "know" is to be taken here, as in some other contexts, as acknowledgement of sovereignty, then it is an easy turn that "not know" means a disregard of proper sovereignty, both a failure to rely upon and a failure to respond in obedience.[15] This negative nuance for Israel is the one that is important for what follows in the speeches of judgment in Isaiah. The positive dimension of knowing, here enacted by others of God's creatures, is knowledge of God's rule and God's ordering of creation for the sake of food. This is the most elemental knowledge required, a knowledge available to and practiced by all of Yhwh's creatures, with the conspicuous exception of Israel, the only creature of the creator on the horizon of this brief poetic unit who does not know.

V.

The second text I mention is Jeremiah 8:7:

> Even the stork in the heavens
>   knows its times;
> and the turtledove, swallow, and crane
>   observe the time of their coming;
> but my people do not know
>   the ordinance of the Lord.

The content of the verse is closely paralleled to that of Isaiah 1:3. Again the contrast is between "my people" and the nonhuman creatures, conventionally translated as storks, turtledoves, swallows, and cranes. Again the contrast is a sharp one between *knowing* and *not knowing.* The birds know all they need to know; they know the seasons of the creator:

---

[15] See H. B. Huffmon, "The Treaty Background of Hebrew *Yada⋅*," *BASOR* 181 (1966) 31–37, H. B. Huffmon and S. B. Parker, "A Further Note on the Treaty Background of Hebrew *Yada⋅*," *BASOR* 184 (1966) 36–38.

> seedtime and harvest,
> cold and heat,
> summer and winter (Gen 8:22).

As the poetry of Isaiah 1:3 is verified by the regular phenomenon of cattle headed home for feeding at the right time, so the poetry of this verse is verified by the migrating patterns of birds, most spectacularly by the swallows of Capistrano. The birds know how the world is ordered by the creator, and thus they "observe the time" which is to be in sync with that ordering.

The contrast is parallel to that of Isaiah 1:3. "My people" is the only creature who does *not know* about coming and going. The preceding verses (Jer 8:4-6) have spoken of "turning" and "returning," of going away and coming back in repentance. As the birds know, there are seasons for coming and for going; but "my people" have no sense of season or of appropriate action for a particular season in the ordering of the creator. In the second case as in the first case, Israel's capacity for "knowing" is as easy and available as it is for every other creature. Such knowing requires only to attend to the patterned life willed by the creator. To be "dumber than an ox," to know less than the flight schedule of birds, is not an act of stupidity. It is, rather, an act of deliberate recalcitrance, a defiance that sets the creature over against the creator.

VI.

The force of these poetic images depends upon creaturely knowledge that is God-given and available to all creatures. That is the premise of being a workable creature of God. This knowledge is not "primitive" or "savage," but it is the premise of responsive creatureliness that adheres to the uncompromising reality of the creator. This is the same knowledge possessed by the "Leavers" that Quinn exposes; all creatures have access to it. With this claim, the prophetic indictment of "not knowing" becomes ominous indeed.

Now of course these poetic passages are not interested in birds or beasts of burden. These are simply foils for the indictment of "my people" who does not know. It is of immense importance, in any case, that the knowing for which Israel has a dangerous deficit is not from Torah or from Sinai revelation. This knowing is given more elementally, more bodily, more creaturely. The *not knowing* for which Israel is condemned is a deliberate act of alienation from the creator, an act of autonomy and hubris that refuses to order life according to the intention of the creator, an attempt to order and control life outside "the hands of the gods."

These prophetic verses, when placed in the context of the contemporary proposals of Quinn and Gabel suggest that "creation theology" is not about nature worship or any of these matters that the early Barth feared. It is, rather, about the elemental awareness given to every creature, human and nonhuman, that creatures must adhere to the givens, limits, and requirements of the creator.

Now, of course, this alternative way of knowing has long been recognized in Old Testament studies. We may take William Foxwell Albright as a striking example of the way in which this alternative knowing has been widely understood in an evolutionary scheme whereby knowing has developed from the "dynamism" of primitiveness to modern rationality:

> Levy-Bruhl stresses the prelogical character of primitive thought, which fails to take account of contradictions, lacks any clear concept of causal relations, for which it substitutes simple explanation by sequence, or superficial concomitance, or accidental resemblance. Fundamental to primitive thinking are also impersonality and fluidity.[16]

Clearly Albright intends to show that such a way of knowing contrasts such knowing with modern knowing, even if the older knowing is not as "primitive" as had been thought. The contrast is inescapable. When placed in an evolutionary scheme as Albright does, however, the "primitive" is inferior and the modern rationality is welcomed as being superior and as the ultimate arrival in epistemology. If, however, we take the contrast of these two modes of knowledge out of an evolutionary scheme and see them as alternative choices, then we may see that the knowledge championed in these texts is a subversive alternative to autonomous rationality. It is my impression that Quinn and Gabel focus upon this subversive alternative knowing because they observe how the autonomous reason of the Takers is death-dealing. It will be an important shift in thinking about such creaturely knowing to see that it is not *an inferior way* of knowledge to be overcome but it *is an alternative* to be embraced, an alternative clearly championed by these two poets in a society that knew everything except what needs to be known about the creator.

I write this as the United States just now strides the earth with its immense, unrivaled technical capacity, sorting out every other culture according to the defining reality of U.S. military and economic power.

---

[16] Albright, *From Stone Age to Christianity*, 169.

The old truth from these poets keeps emerging again and again, how to know so much but not to know the things that make for peace (Luke 19:42). Asses and oxen show up to be fed; birds fly on schedule. But the critically competent do not know, and choose all too often another knowledge that leads to death. It occurs to me that the emergence of the knowledge of wisdom-creation is not just an interesting fresh scholarly trope in Old Testament studies. It is, rather, a resilient alternative that not all the Takers in the world can effectively obliterate. "Not knowing" has no adequate compensation in knowing everything else.

2

Feasting and Forsaking Ignorance

(Proverbs 9:5f.)

RICHARD J. SKLBA

Whoever fears the Lord will do this;
    whoever is practiced in the law will come to wisdom.
Motherlike she will meet him,
    like a young bride she will embrace him,
Nourish him with the bread of understanding,
    and give him the water of learning to drink (Sir 15:1-3).

The biblical image of personified Wisdom as mother and bride offering food and refreshment to any serious student presents a striking
picture. This text also provides a helpful introduction to a theme which
runs throughout much of the biblical testimony over the centuries,
namely, the nourishment which true wisdom serves any individual who
finds a place at her table. No one who partakes with the listening heart
requested by Solomon (1 Kgs 3:9) can leave that repast unchanged. It
is the theme of inner transformation and conversion as experienced at
table which I wish to explore in this brief study. The thesis of the following essay is that table transformation is celebrated in both the Wisdom traditions of Israel and the Gospel of Luke.

THE WISDOM OF JESUS BEN SIRACH

By the beginning of the second century B.C.E. when Jesus Ben Sirach compiled his collection of admonitions for the student of true Wisdom, a long and venerable tradition had preceded him.[1] Written, it would seem, shortly before the persecutions occasioned by the brutal decrees of King Antiochus Epiphanes IV and the subsequent resurgence of the Maccabees (168 B.C.E.), the instructions of Ben Sirach reflected the intellectual accomplishments of a society marinated for centuries in Semitic cultures and further seasoned by the spreading Hellenism of that age.[2] Military revolt and Maccabean victory produced a new sense of pride in the cultural and religious heritage which had blessed early Judaism. At that time a fresh appreciation for Israel's religious values marked the lives of those who fought against rank Hellenism and who reembraced the faith of their ancestors. Wisdom and the sages who transmitted that treasure across generational boundaries often flourished at precisely those times of great cultural change when the religious ideas of the past no longer seemed to hold the loyalty of newer generations.[3] Thus, sometime after 132 B.C.E. the grandson of Ben Sirach translated his grandfather's literary work from Hebrew into the Alexandrian Greek of his day so that his cultured contemporaries in Egypt might also savor the religious heritage of past generations of Israelites.[4]

Long influenced by the traditions of Egypt and other Near Eastern cultures, the wisdom of temple and court scribes in Jerusalem had become a treasure for all of Israel and Judah. Merchants and diplomats visited the various Near Eastern royal courts over the centuries, bringing tales of splendor experienced elsewhere. Entertainment at banquets in Jerusalem, reflecting the oriental mores of that world, undoubtedly

---

[1] See Patrick Skehan and Alexander DiLella, O.F.M., *The Wisdom of Ben Sira* (New York: Doubleday, 1987); and Alexander DiLella, "The Wisdom of Ben Sira: Resources and Recent Research," *Currents in Research: Biblical Studies* 4 (1996) 161–81.

[2] See John J. Collins, *Jewish Wisdom in the Hellenistic Age* (Louisville: Westminster John Knox, 1997). A related and vexing question for students of early Judaism: did the rabbinic stress on study of the Torah result from the influence of ancient sages, or was that practice more accurately understood as the contribution of Hellenism?

[3] Thus the sages seemed to have flourished at four turning points in Israel's history: the introduction of the monarchy under David and Solomon, the new consolidation of faith in Jerusalem at the time of Hezekiah after the destruction of Samaria, the return from Exile, and, finally, the Hellenistic/Roman occupation of Judaea.

[4] The preface to that translation is often included in the introduction to Ben Sirach's work as testimony to its literary provenance and history.

included both music and verbal wit. Gradually a common set of obser-vations on the irony and vagaries of human behavior was shared among the nations linked by commerce, intermarriage, military ventures, or political alliances. The suggestion of this essay, namely, a meal, whether festive or familial in character, for entertaining and instructive ex-changes seems an obvious venue for such reports.

Clever commentary on human behavior as observed by the sages over the centuries was gathered. Caricatures of covert or gross gluttony, human stupidity, and folly became carefully turned phrases. These were brilliantly crafted if occasionally biting portraits of the types of person-alities who paraded across life's stage in every generation. Individual aphorisms gradually became collections of various lengths. Anyone who recognized him/herself in those satires was inspired to change outer behaviors if not inner attitudes. Even the most secular of wisdom engenders conversion at some level.[5] Bits of wisdom gained from bitter human experience could find ready expression when the hours stretched late, when the wine carafe emptied, and when a melancholy mood settled over the banqueters, as reflected, for example, in the sum-mation provided by the comment: "in much wisdom there is much sor-row."[6] At least, that is an imaginary social context which could so easily generate such observations.

These aphorisms and collections also served at times as texts for the education of young nobility just as they had since the courts of David and Solomon. Granted, the only reference to an actual school in the en-tire Old Testament corpus is in fact found in Sirach (51:23).[7] Admoni-tions, however, on the proper etiquette to be observed when invited to festive meals were included in Ben Sirach's collection, together with such basic common sense stipulations as not being the first to reach out when many are at table (31:18). Blessings were declared on the man who was generous with food (31:23). There were clever queries and rhe-torical questions regarding whether a person really lives who lacks wine

---

[5] One immediately thinks of the contributions of Bernard Lonergan regarding the in-tellectual, moral, and religious levels of conversion resulting from knowledge; cf. *Method in Theology* (New York: Herder and Herder, 1972) 241.

[6] NAB Qoh 1:18 which is more poignantly translated by CEV's "The more you know, the more it hurts."

[7] See James Crenshaw, *Education in Ancient Israel: Across the Deadening Silence* (New York: Doubleday, 1998). A very different approach can be found in Donn Morgan's *The Making of Sages: Biblical Wisdom and Contemporary Culture* (Harrisburg: Trinity Press In-ternational, 2002).

(31:27). A rather complete guide was offered by Ben Sirach for anyone presiding at a festive meal (32:1-13). Such a host was advised to have only individuals who were truly just as table companions (9:16). These maxims would have been part of the material commended for rote learning by young students practicing their writing on tablets or engaged in their first attempts at oral declamation.[8]

One is quick to understand the ease with which imagery of banqueting and feasting became associated with the accumulation of wisdom. Wisdom, said Ben Sirach, inebriates and fills the house with choice foods (1:14f.). The pleasure of tasting sweetmeats seemed very similar to the feel of smooth and witty words on the tongue. They provided satisfaction and nourished daily activity with new energy. The occasion for their exchange as well as their content inevitably conjured up images of mealtime.

PROVERBS

The comments of Ben Sirach developed and mirrored the observations of earlier biblical collections. The preface to the book of Proverbs, for example, also spoke of Wisdom as nourishment.[9] Written, it would seem, sometime during the fifth century B.C.E., the nine introductory chapters to a similarly disparate collection of more ancient proverbs also provide advice for people seeking Wisdom, gradually introducing more explicit references to the reality of God within all of human experience.

From the beginning, wisdom had been associated with the practical ability to accomplish something successfully. The artisan was the first labeled as "wise."[10] In later years the concept came to be associated with the ability to negotiate life's challenges in a fashion which elicited admiration from colleagues. The sovereign action of God was gradually acknowledged as the ultimate source for that skill.

[8] Although from a totally different culture and era, the famous *McGuffey Readers* provided an early American version of the use of proverbs and aphorisms in the education of the young. It should be noted, however, that most ancient education of the young was oral. The level of noise generated by a classroom or its ancient counterpart was the reason for the Talmud's requirement of the neighbors' permission prior to the establishment of a school.

[9] See Michael V. Fox, *Proverbs 1–9* (New York: Doubleday, 2000).

[10] Hiram the brassworker from Tyre who worked on Solomon's Temple was filled with wisdom and understanding and skill (1 Kgs 7:14), and the workmen, stonecutters, masons, and carpenters brought together for the construction of that Temple under Solomon were judged skillful, literally "wise," in their craft (1 Chr 22:15).

The earlier assorted proverbs and pithy sayings included references to the folly of foolish behavior and the blessings of wisdom for the thoughtful person. One was advised not to consort with winebibers (Prov 23:20), and Wisdom was described as likened to honey for one's soul (24:13). Being attentive to one's master was recommended as akin to eating fruit from the fig tree (27:18).

These various bits of truth were introduced by a long preface contrasting Dame Folly, whose adulterous invitations led only to destruction, with the figure of Lady Wisdom who invited the simple to sup at her table. Evil people are described as eating the bread of wickedness and drinking the wine of violence (4:17), thus becoming what they consume. By contrast Wisdom spreads a lavish table and invites people to enjoy the repast. As a figure of personified hospitality and generosity, Wisdom is described as graciously nourishing human beings who come humbly to her home.

Once again, this preface was written at an age of change, just subsequent to the return from Babylonian Exile and the rebuilding of the Jerusalem temple. Although still under the domination of the Persian Empire at the time, a sense of pride over their preserved heritage also marked the early fifth century B.C.E. just as it would characterize the later period of Jesus Ben Sirach.

Often during times of religious transition, especially periods of profound social change when the religious and theological tenets of earlier ages seemed less capable of supporting the entire freight of the religious tradition, the sages fell back upon their human experience and its lessons.[11]

In those earliest times, and again in the periods of theological change, it would seem that the recognition of divine activity remained in the background of this literature. Only later did one find in the writings of sages from Israel and Judah a more explicit identification of Wisdom with personified revelation and eventually with Torah (Sirach 24).

At some point the food and drink served by Wisdom became an allegory for nourishment and growth as indicated in the brief reference/contrast with Dame Folly at the end of the Prologue (Prov 9:5f.). As always, feasting remained the privileged time for an entertaining exchange of ideas, and for the instruction of younger people who were being groomed for future leadership. As the conversation unfolded, everyone was changed.

[11] For an excellent contemporary introduction, see Dianne Bergant, *Israel's Wisdom Literature: A Liberation-Critical Reading* (Minneapolis: Fortress, 1997).

Teaching and instruction, however, were not restricted to the lives and work of the sages in Israel. The prophets also found it helpful at times to describe their own initial summons and messages as food. The visions of Ezekiel included the symbolic eating of the scroll covered with writing front and back, inscribed "lamentation, wailing and woe" (Ezek 2:10). It tasted like honey in his mouth (3:2).[12]

MEALS IN ISRAEL

A popularized though classic description of the foods in an ordinary family's diet can be found in Daniel Rops's *Daily Life in Palestine at the Time of Christ*.[13] The primary focus of that section of his study was the description of the foods actually consumed, rather than the social institutions or customs surrounding that human activity. Curiously, when Roland de Vaux, o.p., published his *Les Institutions de l'Ancen Testament* in 1958 and 1960, any references to meals were only presented by reason of their relation to cultic ritual and sacrifices.[14] No concern seemed evident for other categories of social dining. Except for the context of mourning or worship, references to the staples of life and meals were not on de Vaux's screen.

By contrast, *Life in Biblical Israel*, the recent volume of Philip J. King and Lawrence E. Stager, incorporated more information regarding meals in daily life. This change in scope may have been inspired by developments in the science itself and in the range of newer archaeological interests over the past half century. As different scientific specializations became increasingly frequent among the abilities of excavation members, each new set of eyes found more information. The result was a deeper knowledge of the daily life and eating habits of the people who once lived amid the walls and lanes of ancient cities. Elaborate hospitality toward guests was a duty and a privilege. Common meals reinforced family bonds and strengthened the relationships among people.[15]

---

[12] The book of Revelation takes up the same prophetic theme as the seer is commanded to consume the scroll which tasted like honey in his mouth, but turned sour in his stomach (Rev 10:9f.). A similar intermingling of themes is found in the sixth chapter of John's Gospel where revelation and Eucharist are interwoven under the theme of "bread from heaven."

[13] (London: Weidenfeld & Nicolson, 1961) 199–211.

[14] The English trans. by John McHugh was published as *Ancient Israel: Its Life and Institutions* (London: Darton, Longman and Todd, 1961).

[15] (Louisville: Westminster John Knox Press, 2001) 61–68.

MEALS IN THE GOSPEL OF LUKE

Still more information on this subject was provided by the introductory sections of the volume by Eugene LaVerdiere, s.s.s., entitled *Dining in the Kingdom of God: The Origins of the Eucharist according to Luke*.[16] Father LaVerdiere made a helpful distinction between two types of meals in first-century Judaea. Thus in the categories described by LaVerdiere's study of Luke there is, first, the traditional meal of Semitic hospitality known since before the days of Abraham and Sarah who entertained the three guest angel messengers of God (Gen 18:5).[17] The conversation during such meals would have been more familial, but also instructive as the wisdom of divine messengers or elder family members was "served" together with the staples of physical nourishment and the social exchange of daily life.

There was also the type of meal which LaVerdiere calls the *sumposion* in Hellenistic society.[18] Such a banquet would have presumed guests of considerable affluence gathered in a space ample enough for dining couches, either in the more ancient Hellenistic arrangement of individual couches, or in the later "modern" Roman style of dining couches large enough for three guests each. In this case, the host or a specially honored guest would provide the entertainment expected by the sophisticated participants. Wisdom was thus served as nourishment.

Luke describes ten accounts of meals at which Jesus the Teacher is present. Woven through each meal description, of course, are the subthemes familiar to students of Lukan theology: the call to conversion, a perennial welcome to sinners, care for the needy and outcasts, the presence of women as disciples/hostesses and colleagues, the quality of hospitality expected of a true disciple, the value of service, and the necessity of accepting one's cross in order to enter the Kingdom of God.[19]

---

[16] (Chicago: Liturgy Training Publications, 1994).

[17] The five Lukan meals which LaVerdiere placed in this Semitic category were the picnic meal/multiplication (9:10-17), Martha's dinner (10:38-42), Zacchaeus's feast (19:1-10), and the two post-resurrection meals (24:13-35 and 36-49).

[18] These five in LaVerdiere's division would include Levi's banquet (5:27-39), the three dinners at the home of a Pharisee (7:36-50; 11:37-54; and 14:1-35), and the Last Supper/Passover (22:14-38).

[19] There are other references to food in Luke's Gospel, but they are not descriptions of true meals, as for example, the plucking of grain on the Sabbath (6:1ff.), the beatitude's blessing of ultimate reversal for the hungry (6:20-25), life itself being more important than food (12:23), the missionary mandate to eat whatever is served while announcing the Kingdom of God (10:7, which is clearly an occasion for conversion), the humility of seeking

These ten meals occur at significant moments in the journey of Jesus from his baptism in the Jordan to his passion in Jerusalem. Although the Lukan Jesus is primarily cast in the mantle of the Elijah and Elisha traditions, he also travels as a sage, offering his instruction to disciples of his choice. The wisdom themes in these stories include generous hospitality to the humble, the condemnation of arrogance, blessings for the industrious, the benefits offered to those who are open to instruction from others, and the rewards of living a just and virtuous life.

At meals surrounded by friends and foes, often at the invitation of Pharisees who extended hospitality in order to engage him in conversation about the Reign of God, Jesus gave teachings and expounded his call to *metanoia* and renewal of life. His instruction illustrated the path to full happiness and holiness.

The first mention of food in Luke (4:1-4) is a negative story about abstention from physical nourishment in a time of fasting and testing. Perhaps there remains a subtle allusion to the hungers of Israel after leaving the plentiful lands of Egypt, preparing for greater dependence on divine hospitality and the daily gift of manna. The hunger becomes an occasion for learning about other needs even more fundamental than bread alone. That is also wisdom.

The ten genuine meal stories in Luke seem to be occasions for Jesus to share his divine wisdom and opportunities for conversion, or at times descriptions of such graces rejected. Perhaps a brief recounting of the conversion meals in Luke will enable readers to trace the theme more precisely.

Thus the second mention of food is the banquet after the call of Matthew/Levi, the tax collector (5:27-32). This becomes the *first true meal* account in Luke. Some of the Pharisees and scribes complained to the disciples about their eating and drinking with tax collectors and sinners.[20] The response of Jesus the Sage was the famous admonition of the need

---

the lowest place at a wedding feast (14:8), the parable reference to the father's festive celebration of the return of the prodigal son (15:23), and parable's condemnation of the inability of the rich man to see the hungry beggar at his doorstep while he feasted splendidly (16:19). A classic summary of Lukan scholarship can be found in the two volume Anchor Bible Commentary of Joseph Fitzmyer's *The Gospel according to Luke I-IX* and *X-XXIV* (New York: Doubleday, 1981 and 1985). See also J. Fitzmyer, *Luke the Theologian: Aspects of His Teaching* (New York: Paulist, 1989).

[20] Tax collectors were already welcomed by John the Baptizer when he counseled justice but did not demand their resignation (3:12), thus demonstrating a certain pre-evangelical welcome. Wisdom was at work in his preaching as well.

of the sick for physicians, not the healthy. In this manner the Gospel begins to establish meal stories as occasions for conversation with an eye toward inner healing, *metanoia*,[21] and conversion. This change of heart, or movement toward thinking in a new way about the mysterious works of God in the world, is the fundamental purpose for the presence of Jesus at Levi's home. Whether this be a festive Hellenistic *sumposion* or the more Semitic friendly gathering, the account is thus linked in concept and purpose to the wisdom meals of ancient Israel. This account of Levi's celebration will also be echoed in the feast of celebration hosted by Zacchaeus, another tax collector at the end of Luke's account of the journey (19:1-10). Wisdom is clearly at work in both instances.

The *second meal* account occurs at the home of Simon the Pharisee where the meal is interrupted by the visit of the sinful woman (7:36-50). In bathing the feet of Jesus with her tears as she kissed and anointed them, she herself experienced the conversion of such a meal, but not the host who entertained in silent judgment. Simon was not open to a "torah" of welcome to sinners also journeying toward divine forgiveness and salvation. Wisdom was rejected by Simon who separated himself from sinners and lost the lessons their lives might offer.

Shortly after the selection of the Twelve is Luke's account of the multiplication of the loaves near Bethsaida (9:10-17). In this picnic pericope it is precisely the Twelve who ask Jesus to send the crowd away, claiming to have only five loaves and two fish. It is the Twelve in turn who are instructed to settle the crowd and distribute the food. Because it is precisely twelve baskets of leftovers and fragments which remain, the point of the story must be aimed at them. Whatever their limitations, this *third meal* account insists that they must come to learn that their resources will be sufficient because God will supply for the human deficiencies. Wisdom was served together with the fish.

The *fourth meal* account mentioned by Luke is the dinner provided by Martha and Mary (10:38-42). A change in thinking was demanded by the behavior of Mary. The instruction of Jesus is that the posture of discipleship is even more important than the traditional duties of generous hospitality. The customary welcome had been mandatory since before the time of Abraham and Sarah who hosted the three visiting angels at Mamre (Gen 18:1-15). In the conversation of Jesus with Martha an even higher measure of conduct was taught. The real main course of that meal was Wisdom about entering the Kingdom of God.

---

[21] Literally, "thinking outside one's own mind."

The *fifth meal* story occurs on the occasion of hospitality extended by yet another (unnamed) Pharisee who observed that Jesus had entered his home and reclined at table without the customary ablutions (11:37-52). The ensuing conversation became a strong call for interior conversion. Tithes on herbs, intended to reflect gratitude for even the smallest of blessings, were acknowledged as obligatory, but even moreso was Wisdom's demand for justice and love of God. Into that meal story Luke chose to insert the Lord's strong condemnations of the sins peculiar to the experts in Torah. Perhaps the story was an account of conversion rejected by the legal scholars, or at least without any external sign of changed behavior or thinking on their part.

The *sixth meal* occurs at the home of one of the leading Pharisees (14:1-24). The Sabbath cure of the man suffering from dropsy became an opportunity to teach the preeminence of responding to human needs over the requirement of abstaining from work on that day of rest. It also became the context selected by Luke for a teaching about humility and not seeking places of honor at table. Ben Sirach's ancient Wisdom and his admonitions regarding table etiquette found a new home.

The *seventh meal* story also occurred in connection with the story of a tax collector, namely Zacchaeus (19:1-10). The self invitation of Jesus, together with the explanation of his purpose in visiting the home of Zacchaeus, doesn't actually include any description of the banquet itself. The history of interpretation of this text has presented a clearly divided tradition regarding the specific character of the story. Those who claim that the statement of Zacchaeus announces a dramatic conversion to greater generosity and justice in his occupation as tax collector will insist on translating the verb as future, i.e., "I will give" (v. 8). This would then be a conversion story which proceeded to a celebration of the wisdom gained in the exchange. Over the centuries other interpreters have noted the present tense of the verbs in the Greek text *(didomi/apodidomi)* and concluded that Jesus is recognizing Zacchaeus's form of heroic generosity and justice as characteristic of any true son of Abraham. In such a version it is the conversion of everyone else which would be expounded at the banquet in the hope of extending the spiritual wisdom of Zacchaeus to the rest of the participants. Jesus, the divine Sage, points out the true way of successfully maneuvering through life. The way of Wisdom is not by arrogance or greed, but by generosity and justice.

The *eighth meal* is the Lord's final celebration of Passover with his disciples (22:14-38). The profound conversation took the form of teaching about the ultimate blessing of life under the appearance of bread and wine, broken, poured out, and given away. That meal was also the occasion for a final teaching regarding the preeminence of leadership as service. Perhaps this was the quintessential form of the Wisdom offered at table by Jesus the divine Sage. The simple were the most favored guests of Wisdom!

The *ninth meal* in Luke's Gospel is the Risen Lord's memorable and moving supper with the two disciples at Emmaus (24:30-32). The conversion they experienced so profoundly was the realization of his unsuspected risen presence in their midst as the ultimate verification of the teachings of Moses and all the prophets (24:27). Wisdom could be both stunning and matter of fact.

Finally, the *tenth meal* occurred back in the upper room at Jerusalem (24:36-49) where Jesus asked for something to eat as proof of the full reality of his risen presence. That meal became yet another occasion for instructing the astonished disciples about all the teachings of the Torah and the prophets and the psalms regarding his final victory over sin and death. The suffering servant, meek and without deceit, became the key to full and final success in life. In that meal which shared cooked fish, they were declared the official witnesses to his resurrection and were instructed to await the gift of the Spirit so that they might begin their mission of proclaiming conversion to the world. Wisdom would again be proclaimed in the streets and highways of all nations.

Thus, this essay suggests that the ten meal accounts included in the Gospel of Luke become opportunities for conversion welcomed or rejected, and on each occasion Wisdom itself is either welcomed or rejected as well. These reflections are offered in personal tribute to Diane Bergant, whose every lecture has always been the serving of the type of Wisdom which truly nourishes and compels change in our minds and hearts.

Indeed, "Happy the one who eats bread in the Kingdom of God" (14:15).

3

Redeeming "Gospel Feminism" from Anti-Judaism

MARY C. BOYS, S.N.J.M.

Since this essay appears in a *Festschrift* honoring Dianne Bergant, I begin on a personal note. Unlike most of the other authors in this volume, I do not know Dianne well, and have never worked in geographic proximity to her. Yet my first sustained encounter with her was so powerful that it provides a lasting bond. In July 1998 Dianne and I made retreat together in Monroe, Michigan, with ten other women theologians belonging to religious communities. The level of sharing was extraordinary; the opportunity to engage with her and the others in shared prayer, ritual, and conversation has given me a deep affection and profound respect for Dianne. So it is a privilege to write for this volume on a topic drawn from the wellspring of our shared love of the Scriptures and shaped by our passionate feminist perspectives.

THE QUESTION

Christian feminists, in attempting to reveal the liberating power of the Way of Jesus Christ for women in our time, may inadvertently continue the transmission of a theological "virus" that has tragically affected Jews and distorted the self-understanding of Christians:

anti-Judaism. My principal concern is to analyze ways in which the insidious virus that misrepresents Judaism has managed to attach itself to biblical interpretations directed toward justice for women in our Church and world.

The question is this: How might Christians who advocate for greater justice for women also contribute to reconciliation with Jews? How can Christian feminists eradicate the virus of anti-Judaism?

CLARIFYING TWO TERMS:
ANTI-JUDAISM AND GOSPEL FEMINISM

The title of this essay rests on a distinction between anti-Judaism as a theological viewpoint and antisemitism as a socio-cultural phenomenon. By anti-Judaism I mean attitudes, arguments, polemics, and actions that distort and disparage Judaism in order to support Christian claims of superiority. This is a complicated distinction, since understandings of Christianity resting on anti-Jewish perspectives can—and do—all too readily "slide over" into antisemitism, that is, into hatred of and hostility toward Jews.

Moreover, anti-Judaism is not a univocal term. It has taken on new layers over time, becoming ever more deadly. What initially grew out of an argument over differences between those who believed in Jesus (including Jewish disciples) and (other) Jews grew, over the course of many centuries, into a bitter dispute resulting in harsh treatment of Jews. The Church consistently rationalized its actions by defaming Jews as a ",deicide people" and "Christ-killers." At the heart of anti-Judaism is the charge echoed repeatedly that the Jews bear the burden of responsibility for the death of Jesus. Its nearly constant accompaniment is the caricature of Judaism as legalistic, personified by the hypocritical Pharisees. Over the ages, anti-Judaism fueled hostile attitudes toward Jews that in many cases gave rise to persecutions and even death.

As reprehensible as this legacy is, it is not identical with antisemitism, nor is antisemitism the only cause of the *Shoah* (Holocaust). Nevertheless, the Shoah demands examination. For Christians, in particular, it requires confrontation with the lethal legacy of anti-Judaism—a confrontation that the Catholic Church, among others, takes very seriously in our time.

Because anti-Judaism has replicated itself in many dimensions of Christian theological thinking for nearly two thousand years, it will be neither neatly nor quickly extricated. It is so pervasive that many

Christians of great integrity are unaware of the distorted views of Judaism that they inherited. This does not make them anti-Semites. Thus, in criticizing the theological perspectives of various feminist theologians, I am not accusing them of antisemitism. I am, however, questioning whether they have done justice to the complex relationship of Judaism to Christianity.

A word also about the term "Gospel feminism." I first became conscious of this term while collaborating on a statement with fifteen other women, each of whom had delivered the Madeleva Lecture at Saint Mary's College in Notre Dame, Indiana. The college convened us in late April 2000 to compose a brief "charter" for women. We produced "The Madeleva Manifesto: A Message of Hope and Courage." One of its paragraphs reads:

> To young women looking for models of prophetic leadership, we say: walk with us as we seek to follow the way of Jesus Christ, who inspires our hope and guides our concerns. The Spirit calls us to a gospel feminism that respects the human dignity of all, and who inspires us to be faithful disciples, to stay in the struggle to overcome oppression of all kinds, whether based on gender, sexual orientation, race or class.

I associate the term Gospel feminism especially with the New Testament scholar Sandra Schneiders (who also made our retreat in 1998). In her 2000 Madeleva Lecture, "With Oil in Their Lamps: Faith, Feminism and the Future," she defined feminism as "a comprehensive ideology, rooted in women's experience of sexually-based oppression, that engages in a critique of patriarchy as an essentially dysfunctional system, embraces an alternative vision for humanity and the earth, and actively seeks to bring this vision to realization."[1] Schneiders grounds this alternative vision in the example of Jesus as he is presented in the *Gospel;* she believes his example can contribute to the transformation of *cultural* feminism into a prophetic force that moves all creation toward a future in God's universal shalom.

Whether or not the term originates with her is not relevant; more important, as I shall show later, is how the term functions for those who use it, including Schneiders.

---

[1] Sandra M. Schneiders, *With Oil in Their Lamps: Faith, Feminism and the Future* (New York: Paulist, 2004) 4. Page numbers in further references to this book will be placed in parentheses.

PATRIARCHAL JUDAISM, LIBERATING JESUS

Feminist theology has developed considerable depth and breadth in its brief history. Thus, one encounters significant variations in assumptions, methods, foci, conclusions, and degrees of sophistication. My concern is as much with popular versions as it is with scholarly perspectives, since the former tend to circulate more widely. Especially at the popular level, it is all too likely that the blatant anti-Judaism woven into argumentation will go unnoticed, particularly because the authors champion a good cause—one I, too, hold dear: justice for women in a Church that squanders their gifts and dedication.

A prime case in point appears in the following:

> Palestinian Hebrew women were among the poorest in the world in Jesus' day. This was probably because they had no inheritance rights and could be divorced for the flimsiest of reasons. Hebrew men could divorce their wives for anything from burning the dinner (Hillel) to adultery (Shammai). Yet Hebrew women were not allowed to divorce their husbands. . . . A Hebrew woman had minimal to no property rights . . . . A child was held to be Jewish only if the mother was Jewish. Most Jewish girls were betrothed by their fathers at a young age. Jewish women were held to be unclean while menstruating. If she inadvertently touched a man while having her menses, he was obliged to undergo a weeklong purification ritual before worshipping at the Temple . . . In early Judaism women did proclaim and prophesy, but in Jesus' day they weren't permitted to proclaim Torah at synagogue because of their periodic "uncleanness." As a rule, only the Rabbis' wives were so educated. Women were not accepted as witnesses in Jewish law, nor could they teach the law. Women had no official religious or leadership roles in first century Judaism. Jesus' behavior toward women, even viewed through the androcentric lens of the gospel texts, is remarkable.[2]

We have here a treasure trove of anti-Jewish themes, all revolving around the dominant motif: Judaism was hopelessly patriarchal. Its women were impoverished, subject to the whims of their husbands, forced into marriage at a vulnerable age, excluded because of menstruation, banned from giving witness or teaching, and barred from leadership. Not quite the Taliban—but close.

Portraying Judaism as utterly constricted by patriarchy then allows this writer (and many others) to remove Jesus from his Jewish matrix by asserting that that Jesus' revolutionary attitudes toward women

---

[2] Christine Schenk, c.s.j., "Celebrating the Inclusive Jesus," *Celebration: An Ecumenical Worship Resource* (February 2000) 81–82.

broke the grip of patriarchy. It is this Jesus who shows us what gender relations should be in the Church today.

Consider the following: "At the historical moment when Jesus was born into the world, the status of Jewish women had never been lower. . . . By the time of Jesus' birth, many decades or rabbinic commentary and custom had surrounded Old Testament literature. And these rabbinic traditions considerably lowered the status of women."[3] Or, "Jesus' relations with women seem to have been remarkably free, given the reserve that Jewish custom in his day required."[4]

I could multiply such examples, as they seem ubiquitous in feminist writing, both in the scholarly realm (though with somewhat less frequency now, if more common among feminist authors from the Third World and East) and among pastoral authors. We need in particular to pay attention to what those who write for popular publications are saying, given their wide dissemination.

My encounter with Judaism provides a critical lens on such claims, especially those related to the Second Temple Period and Christian origins. In general, I think much of the anti-Judaism stems from insufficient knowledge of the complexities of this era, and from a general lack of awareness of developments in the dialogue between Jews and Christians.

I am sympathetic to the desire to understand Jesus as one who liberated his followers—including (perhaps especially) women—from the oppressive structures of religious authorities. I too consider some religious authorities oppressive in our time, and wrestle mightily with the harmful consequences of their hubris and abuse of power. Nonetheless, I have come to agree with Mary Rose D'Angelo that portraying Jesus as the one "who saves women from Judaism" both oversimplifies and distorts the situation of Jewish women in first-century Palestine.[5] It does indeed seem, as Amy-Jill Levine writes, that in many feminist and liberationist circles that "no plot is complete without a reference to the 'Jewish

---

[3] Cited in Judith Plaskow, "Feminist Anti-Judaism and the Christian God," *Journal of Feminist Studies in Religion* 7:2 (Fall 1991) 104–07, in Helen P. Fry, *Christian-Jewish Dialogue: A Reader* (Exeter, U.K.: University of Exeter Press, 1996) 233.

[4] Monique Alexandre, "Early Christian Women," in *A History of Women in the West: Vol. 1: From Ancient Goddesses to Christian Saints,* Pauline Schmitt Pantel, ed. (Cambridge: Belknap Press of Harvard University Press, 1992) 420.

[5] Mary Rose D'Angelo, "Gender in the Origins of Christianity: Jewish Hopes and Imperial Exigencies," in *Equal at the Creation: Sexism, Society and Christian Thought,* Joseph Martos and Pierre Hegy, eds. (Toronto: University of Toronto Press, 1998) 25.

patriarchal system,' although very rare are comments on the similarly patriarchal pagan cultures of antiquity."[6] Sadly, I have come to see that many who do theology under the rubric of feminism have drunk all too deeply from the wells of anti-Judaism.

Despite the rich literature on the Pharisees, most notably that of Anthony Saldarini, many Christian feminists continue to regard them as the symbol of precisely what is wrong with religious authorities.[7] For example, one feminist scholar asserts that the Pharisees (and their colleagues, the scribes) "tried to keep control over people's access to God through control of the Law. Jesus denounced this pretension and returned to the people the true means of access to God: love and compassion."[8]

In many feminist writings, moreover, the laws of ritual purity, which the Pharisees are portrayed as championing, serve as the source of the clearest contrast between oppressive Judaism and liberating Jesus. We are instructed about the "dehumanizing situation in which the women of the time were enslaved." We learn how menstruating women were "discriminated against, degraded, and dehumanized."[9] Rather than explore the customs of the Greco-Roman world and probe the mindset of ancient cultures, these authors implicitly blame patriarchy on Judaism.

In short, some feminist writers contrast Jewish women of the Second Temple Period with New Testament narratives to support their claims that "at its inception, pristine Christianity and Jesus himself were free of any misogyny or gender bias." Thereby, as Ross Kraemer notes, they implicitly, even inadvertently, buttress the argument of Christian superiority over Judaism.[10]

---

[6] Amy-Jill Levine, "Lilies of the Field and Wandering Jews: Biblical Scholarship, Women's Roles, and Social Location," in *Transformative Encounters: Jesus and Women Re-Viewed,* Ingrid Rosa Kitzberger, ed. (Leiden: Brill, 1999) 332.

[7] See Anthony J. Saldarini, *Pharisees, Scribes and Sadducees in Palestinian Society* (Wilmington, Del.: Michael Glazier, 1988).

[8] Tereza Cavalcanti, "Jesus, the Penitent Woman, and the Pharisee," *Journal of Hispanic/Latino Theology* 2:1 (1994) 40.

[9] H. Kinukawa, *Women and Jesus in Mark: A Japanese Feminist Perspective* (Maryknoll, N.Y.: Orbis Books, 1994) 12, 27.

[10] See Ross S. Kraemer, "Jewish Women and Christian Women: Some Caveats," in *Women and Christian Origins,* Kraemer and Mary Rose D'Angelo, eds. (Oxford: Oxford University Press, 1999) 36.

JESUS AND HIS JEWISH CONTEMPORARIES

While the virus of anti-Judaism can be readily detected in more popular works, it seems more deeply buried—a default setting—in the work of some of the most respected scholars. I think of Sandra Schneiders, for whom I have the utmost personal and professional regard, yet whose portrayal of Jesus vis-à-vis his Jewish contemporaries concerns me. In the first volume (of three) of her book on religious life in a new millennium, *Finding the Treasure,*[11] she argues cogently about the way in which the prophetic character of religious life places communities in tension with ecclesiastical authorities. I appreciate this argument and agree wholeheartedly. I disagree, however, with claims such as "Jesus was rejected by official Judaism" (334) or the more subtle assertion that "like Jesus who taught both respect for ecclesiastical leaders insofar as they legitimately hold office and resistance to ecclesiastical oppression, Religious must stand with and for the disempowered in the Church" (cf. Matt 23:2-5 [the scathing denunciation of the scribes and Pharisees]) (255). It is simply not historically accurate to regard the Pharisees as representatives of "official" Judaism. There was no such official body. Moreover, scholars increasingly conclude that we must draw conclusions about the nature of the Pharisaic movement with care and modesty. Sources do not suffice to give us a full picture of this complex movement.

Even less should we use Pharisees as symbols of clericalism, however useful that might be to feminist causes. They were not the equivalent of our ecclesiastical leaders—though I must confess the scathing critique of Matthew's Jesus against those who "make their phylacteries broad and their fringes long . . . [who] love to have the place of honor at banquets and the best seats in the synagogues, and to be greeted with respect in the marketplaces, and to have people call them rabbi" has more than once come to mind during certain Christian liturgical settings.

I am also uncomfortable with Schneiders's section on Jesus as a prophetic model in her otherwise superb 2000 Madeleva Lecture. Her argument is this: Christians must incarnate Jesus' prophetic life and mission in our time. This implies being a "locus of the encounter between God and the world," just as Jesus was. There is a tensive quality to this, as Christians are to be neither totally passive (removed from the

---

[11] Sandra M. Schneiders, *Finding the Treasure: Locating Catholic Religious Life in a New Ecclesial and Cultural Context* (New York: Paulist, 2000) 4.

culture) nor unilaterally active (assimilated to the culture). "Genuine, ambiguity-ridden, two-way Gospel-infused cultural participation is something we as a church community have not yet managed to imagine into existence" (97–108).

Schneiders suggests that Christians look to the way in which the Gospels present Jesus as a prophet. She identifies four features characteristic of Jesus' participation in his culture:

1. Jesus gave his life for his fellow human beings without losing himself.
2. Jesus negotiated the tension between his ancestral faith—his religious tradition—and his mystically rooted prophetic spirituality. "He both belonged to and transcended his tradition" (100).
3. Jesus mediated the tension between the particularity of his life and situation and the universality of his concern through the category of the Reign of God.
4. Jesus lived the tension between radically subverting the social, political, and religious status quo and absolutely refusing to use violence to change society.

I believe that readers unfamiliar with the wealth of scholarship on Second Temple Judaism may draw negative conclusions about Judaism from Schneiders's exposition of three of these characteristics, particularly since the Madeleva volumes appeal to a wide audience, not merely to scholars. So I propose a close reading of her exposition of characteristics two through four.

Jesus negotiated the tension between his ancestral faith—his religious tradition—and his mystically rooted prophetic spirituality. Schneiders notes from the outset of this section: "Nowhere in the Gospels do we find Jesus repudiating his Jewish tradition." She then cites references to his circumcision into the covenant, his responses to the Tempter drawn from Deuteronomy, his regular attendance at the synagogue and observance of Passover, his respect for the Law and the quotations from the Psalms as his dying words. "Nevertheless," she writes:

> Jesus was executed by the Romans on charges brought against him by the officials of Judaism. Jesus broke the law of the Sabbath when his prophetic consciousness called him to do good and to save life (see, e.g., Lk 13:1-17). He cleansed the Temple when its guardians made it a den of commerce (Jn 2:13-17). He threatened the Temple when it was invoked as a shield for infidelity to God (cf. Mt 23:37–24:2). He excoriated the religious official who connived with the wealthy and condemned the poor (Mark 12:40-44), who imposed unbearable religious burden upon the already oppressed (Mt 23:4), who condemned those they consider "sinners"

and exalted their own virtue (Luke 18:9-14), who gave themselves religious titles and claimed seats of honor in the religious assembly (cf. Mt 23:5-7), who arrogated to themselves the administration of God's justice (e.g., Jn 7:53–8:11). Jesus did not oppose his personal spirituality to his religious tradition but expressed his spirituality through his religious practice, even as he freely criticized the religious institutions out of his own experience of union with God. No one controlled Jesus' access to and relationship with God, but he was able to make his spirituality a resource for the reform of his tradition rather than as an alternative to it. Jesus, in other words, belonged truly and deeply to his religious tradition but was neither merged with it nor imprisoned by it (pp. 100–01).

This paragraph glosses over the tremendous diversity within first-century Palestinian Judaism. Consider, for example, Jesus' challenge to the officials of the Temple: Whatever the meaning of his overturning the tables of the moneychangers in those pre-Euro days, it appears to be a one-time prophetic act (perhaps with apocalyptic overtones). We have no other evidence that Jesus opposed the Temple as such, nor did he seem preoccupied with taking on its officials. Contrast this with the harsh invective of the covenanteers of Qumran against the wicked priests found in the Dead Sea Scrolls.

We must always take great care with our formulations regarding responsibility for the death of Jesus, since the charge of deicide, initially made by Melito of Sardis in the late second century, lingered for so long in the Church and has had such tragic consequences. Even though Schneiders rightly attributes Jesus' execution to the Romans, to suggest that the "charges brought against him by the officials of Judaism" motivated the Romans lacks nuance. To be fair, it is not easy to sort out the complex historical and legal matters surrounding the Gospel presentations of the death of Jesus. Raymond Brown devotes nearly seventy pages to such questions in the first volume of *The Death of the Messiah*[12] (329–97). He argues that it is a historical oversimplification to deny any Jewish participation in the crucifixion. At the same time, Brown is acutely sensitive to the "harmful way in which the PNs [passion narratives] have been misused against the Jews" (386). Of interest is his observation:

> Historically we know of teachers and leaders in the Judaism of Jesus' time who were genuinely religious. Rather than blaming the authorities, one must reflect more carefully on the reaction produced by Jesus, a sharply

[12] Raymond Brown, *Death of the Messiah: From Gethsemane to the Grave*, 2 vols. (New York: Doubleday, 1994).

challenging figure who could not always have been received sympatheti-
cally even by the truly religious. . . . The Gospel portrait implies that Jesus
would be found guilty by the self-conscious religious majority of any age
and background. More than likely, however, were Jesus to appear in our
time (with his challenge rephrased in terms of contemporary religious
stances) and be arrested and tried again, most of those finding him guilty
would identity themselves as Christian and think they were rejecting an
impostor—someone who claimed to be Jesus but did not fit into their con-
ception of who Jesus Christ was and how he ought to act (392–93).

By implication, then, we cannot simply implicate "religious authorities," whether Jewish or Christian/Catholic.

A third characteristic of the way in which Jesus participated in his
culture according to Schneiders is that he *mediated the tension between
the particularity of his life and situation and the universality of his concern through the category of the Reign of God.* She writes: "We can only
marvel that this utterly particular first-century Jewish man somehow
transcended the narrowness that had marred religion from long before
his birth in the exclusivity of the Jews and has ever since his death and
resurrection in the Crusades and persecutions of Christians" (103–04).

Here we confront a longstanding charge against Jews and Judaism:
their exclusivism has fostered a narrow religion. Schneiders amelio-
rates this somewhat by linking this exclusivist mindset to the Crusades.
Given, however, Christianity's "teaching of contempt" over the ages
(what the bishops of the Netherlands have termed the "catechesis of
vilification"), to charge Judaism with narrowness and exclusivism without qualifying or contextualizing simply adds to the disparagement.

Finally, Schneiders identifies a fourth characteristic: Jesus lived the
tension between radically subverting the social, political, and religious
status quo and absolutely refusing to use violence to change society. I
cite her at length in order to situate this claim:

> Jesus made available a God who cared particularly about those who did
> not matter in the ecclesiastical and political systems, a God who did not
> have to be approached with gifts or through officials, who did not have
> to be placated by ritual observances, who could not be bought and would
> not be managed. Religion in its true sense, the bond between God and
> humans, was at the very heart of Jesus' person and message and perhaps
> that is why he did not have to talk a great deal about it.
>
> Nevertheless, Jesus was so dangerously subversive of the status quo in
> both the religious institution and the sociopolitical establishment that the
> Jewish hierarchy and the Roman power structure colluded in his execution.

He was formally accused of threatening the Temple, which symbolized the whole religious institution, and condemned to death for threatening Roman civil order by occasioning dissent among the people and their leaders and thus civil unrest that could lead to an uprising. In a pattern still often in operation, the state authorities had a vested interest in keeping the collaborating religious hierarchy in control of the people. Jesus, however, told people that they were free and encouraged them to claim that freedom (106–07).

Did first-century Judaism really teach that God could only be approached via the giving of gifts or the mediation of religious officials? Did it really teach that the purpose of religious ritual was to "placate" God? Here rhetorical excess appears to have trumped scholarship, since no primary sources are cited.

Was Jesus "so dangerously subversive of the status quo in both the religious institution and the sociopolitical establishment that the Jewish hierarchy and the Roman power structure colluded in his execution"? This suggests parity between the Jewish hierarchy and the Roman governing structure that belies the historical situation in which imperial Rome controlled the appointment of the high priest. Referring to "the Jewish hierarchy" is also problematic, for two reasons. First, it suggests Judaism in the first century was organized with a body of officials who had the authority to make and enforce norms for the entire community, analogous, for example, to the magisterium of the Catholic Church. We have no evidence to support this, and much to the contrary. Second, to those with feminist sensibilities—the primary audience of readers—"hierarchy" typically has a negative connotation.

In fact, extant sources do not permit us to understand the precise details of how authority was exercised in first-century Jewish life. We do know, however, that some sort of central council functioned in Jerusalem—a Sanhedrin or *boule*—but "its membership, structures and powers are not clear . . . and probably varied with political circumstances. . . ."[13] Thus, we must be far more tentative in speaking about the functioning of "the" Jewish hierarchy.

[13] Anthony J. Saldarini, "Sanhedrin," in *The Anchor Bible Dictionary*, David Noel Freedman et al., eds. (New York: Doubleday, 1992) 5:979.

LIBERATING CHRISTIAN FEMINISM
FROM ANTI-JEWISH READINGS

I propose two modest means of becoming more sensitive to anti-Jewish readings of Jesus and his Renewal Movement. First, feminists at both the popular and scholarly levels need to deepen their knowledge of the pervasiveness of anti-Judaism in Christian tradition and its tragic consequences. Biblical scholars, in particular, need to examine how the interpretation of texts has functioned in thwarting the full flourishing of women's gifts *and adequate understandings of Judaism.*

Second, continued—and increased—collaboration between Jewish and Christian feminists at both scholarly and pastoral levels will immeasurably enhance our understandings of Gospel feminism. We might learn about the purity laws from scholars such as Paula Fredriksen in *Jesus of Nazareth, King of the Jews: A Jewish Life and the Emergence of Christianity.*[14] From Rabbi Elyse Goldstein's *ReVisions: Seeing Torah through a Feminist Lens*[15] we might discover how a Jewish feminist wrestles with the meaning of the purity rules for her own life. We might correct our tendency to engage in rabbinic proof-texting if we paid heed to Ross Kraemer's essay, "Jewish Women and Christian Origins," in the important volume *Women and Christian Origins* she co-edited with Mary Rose D'Angelo.[16] We might hear new depths of interpretation of the Fourth Gospel from Adele Reinhartz in her *Befriending the Beloved Disciple: A Jewish Reading of the Gospel of John.*[17] By using Jewish commentaries as we study Scripture, we might come to more nuanced understandings of ways Jews also struggle with the meaning of problematic texts.

To recognize that "Jewish hopes of a new and transformed world made the movement begun by Jesus and his companions and early Christian communities forums for participation and leadership for women" necessitates refining our argument but in no way diminishes it.[18] To acknowledge the complexity of patriarchy in antiquity as well as in early Christianity entails facing the arrogance of our own limited perspectives.

---

[14] Paula Fredriksen, *Jesus of Nazareth, King of the Jews: A Jewish Life and the Emergence of Christianity* (New York: Knopf, 1999).

[15] Elyse Goldstein, *ReVisions: Seeing Torah Through a Feminist Lens* (Woodstock, Vt.: Jewish Lights, 1998).

[16] See n. 10 above.

[17] Adele Reinhartz, *Befriending the Beloved Disciple: A Jewish Reading of the Gospel of John* (New York: Continuum, 2001).

[18] D'Angelo, "Gender Hopes," 25.

Confronting anti-Judaism forces Christians to redefine our self-understanding. At the heart of that redefinition is how we understand the way of Jesus Christ for our Church and in our world today. We can understand the liberating, salvific power of his mission and ministry without the odious comparison/contrast with Judaism or Jewish authorities. Justice obliges to do this. It is time Christian feminists drew upon the scholarship of Jews in giving voice to our vision of the liberating power of the Gospel.

4

The Divine "I Am"

Wisdom Motifs in the Gospel of John

BARBARA E. BOWE, R.S.C.J.

INTRODUCTION

Like the figure of Woman Wisdom herself, Dianne Bergant has spent much of her scholarly career inviting others to "Come" and to learn the mysteries of God revealed through Wisdom's teaching. Job, the Song of Songs, Proverbs and Psalms, Ecclesiastes, the Wisdom of Solomon, and Sirach—each of these texts has received careful scrutiny by this teacher of wisdom, and each wisdom book has become more intelligible to readers because of her guiding hand. Summarizing for a popular audience the final influence of Wisdom in the Bible, however, Bergant wrote the following: "The figure of Woman Wisdom, the one who was with God at creation and who pitched a tent among a chosen people, influenced the way early Christian writers described Jesus, who 'was in the beginning with God' and who 'made his dwelling among us'" (John 1:2, 14).[1] Faithful to this insight, therefore, I offer in gratitude this brief reflection on wisdom motifs in the Gospel of John.

---

[1] Dianne Bergant, *People of the Covenant. An Invitation to the Old Testament* (Franklin, Wis.: Sheed & Ward, 2001) 124.

Because the general topic of biblical wisdom is so vast in scope and its influence on the Gospel of John so profound, I shall limit my comments to three dimensions only. First, I shall review the variety of wisdom motifs evident in this Gospel. Second, I examine the unique Johannine use of the "I am" self-pronouncements on the lips of Jesus. And finally I shall propose some concluding remarks on the wisdom motifs as the key to Johannine Christology.

Before turning to these matters, however, it is important to define what is meant by "wisdom/Wisdom" within the biblical tradition. The biblical books generally designated as the "wisdom literature" (Proverbs, Job, Ecclesiastes, Song of Songs, Sirach, the Wisdom of Solomon, and some of the Psalms), despite their great variety, share a common fascination with wisdom. "The Hebrew term for wisdom *(chokma)* occurs in one form or another 318 times in the OT, and over half of these (183) are found in Proverbs, Job, and Ecclesiastes. Hence these three books, along with Sirach and the Wisdom of Solomon (in these two apocryphal books forms of *sophos/sophia* occur over 100 times), have come to represent Israel's 'Wisdom Literature.'"[2] In these texts "wisdom" describes a particular worldview, a way of living that seeks to know the mysteries of life associated with the divine presence in the world and to choose a path in life that will be faithful to God's ways.

Unlike the historical books (the Law and the Prophets), the wisdom literature shows little or no interest in Israel as a nation. Nor does it rehearse the great acts of God in Israel's history such as the Exodus, the giving of the Torah on Sinai, the establishment of the monarchy and the Temple cult. Instead, the wisdom tradition finds God revealed at every moment at the heart of the world, in the ordinary daily common sense of proverbial wisdom, and in the midst of the great mysteries of life, such as suffering and death that touch all peoples everywhere. Wisdom focuses on the whole of God's creation and is therefore more "cosmopolitan" and "ecumenical" than the story of Israel's salvation history that dominates in the Law and the Prophets.

A particular development of wisdom in Israel, and one that had the most profound influence on the early Christian understanding of Jesus, concerns the way in which Israel's religious imagination personified the character of wisdom. And because the Hebrew term for wisdom *(chokma)* as well as the Greek of the LXX *(sophia)* are grammatically feminine nouns, "Wisdom" was personified as a woman. Not only are

---

[2] Roland E. Murphy, "Wisdom in the Old Testament," *Anchor Bible Dictionary*, 6:920.

her attributes praised by the sacred authors, but she herself "speaks" in the text with a voice that reveals to the world the mysteries of God. These features of the wisdom tradition became arguably the most fertile resource for Christian reflection on the nature and person of Jesus. And it was the Johannine community and its gospel that led the way.

WISDOM IN JOHN

In its presentation of Jesus, no gospel story has been influenced to a greater degree by the language, the motifs, and the theological character of the Israelite Wisdom tradition than has the Gospel of John. In fact, as many Johannine interpreters have repeatedly argued, John's portrait of Jesus is the story *par excellence* of the incarnation of Wisdom. In his classic (1966) and still masterful commentary on the Gospel of John, Raymond Brown identified twelve different "wisdom motifs" in John that, taken together, constitute the unique Johannine portrait of Jesus.[3] A review of these motifs will illustrate Wisdom's pervasive influence on this latest of the four Gospels.

From the opening prologue (John 1:1-18) John's story of Jesus is set in cosmic perspective. It begins not with the traditional opening of the Markan story, with Jesus' encounter with John the Baptist, but in the heavens, with God . . . "in the beginning" (John 1:1). From this opening verse it is clear that the fourth evangelist sees the life of Jesus as the story of one who came "from above." John's depiction of Jesus as the pre-existent Word of God, the *logos/sophia*, is the first wisdom motif identified by Brown.[4] Just as Lady Wisdom existed with God in the beginning (Prov 8:22-33; Sir 24:3-7, 9; WisSol 6:22) and shared the mysteries of the divine realm, so Jesus as the Word was "in the beginning with God" (John 1:2), the one who shared in God's glory "before the world began" (John 17:5). In Proverbs, Wisdom tells her own, similar story: "From of old I was poured forth, at the first, before the earth" (Prov 8:23). The pre-existence of Wisdom, therefore, was the principal interpretive clue employed by John to tell his gospel story of Jesus.

A second motif is the claim that Wisdom shares perfectly in the divine character. In the words of the Wisdom of Solomon (7:25-26), "For she is an aura of the might of God and a pure effusion of the glory of

---

[3] Raymond E. Brown, *The Gospel according to John,* vol. 1 (New York: Doubleday, 1966) cxxii–cxxv.

4 Ibid., cxxiii.

the Almighty; . . . she is the refulgence of eternal light, the spotless mirror of the power of God, the image of [God's] goodness." To a Jewish mind for whom God's oneness was absolute, this description was as close as one could come to ascribing "divine properties" to a being other than God without falling into blasphemy. Through this Wisdom lens, then, we can better understand John's claims about Jesus as the word made flesh: ". . . we saw his glory, glory as of the Father's only Son, full of grace and truth" (John 1:14).

Third, the wisdom tradition employs the metaphor of light to describe one of Wisdom's essential characteristics. In the text of the Wisdom of Solomon quoted above, for example, Wisdom is the "refulgence of eternal light" (7:26). Again in the same text, ". . . she is fairer than the sun. . . . Compared to light, she takes precedence" (WisSol 7:29). Likewise, in the prologue of John's Gospel, the author takes up this theme to describe the Word: "What came to be through him was life, and this life was the light of the human race" (John 1:3-4). And John the Baptist heralds the coming of Jesus with these words: "The true light, which enlightens everyone, was coming into the world" (John 1:9). It is no surprise to hear the Johannine Jesus claim to be "the light of the world" (John 8:12; 9:5) and to see his healing of the blind man as concrete proof of his capacity to be light in the midst of darkness (John 9).

A fourth wisdom motif that has influenced the portrait of Jesus in John's Gospel is the claim that Wisdom descended from heaven to dwell with human beings on the earth, and in particular with Israel. Especially important in this regard is the text of Sirach:

> Then the Creator of all gave me a [his] command, and he who formed me chose the spot for my tent, saying "In Jacob make your dwelling (*kataskenoo* = pitch a tent), in Israel your inheritance" (Sir 24:8).

When the prologue of John claims that "the Word became flesh and made his dwelling (*skenoo* = pitch a tent) among us" this is surely an unmistakable borrowing from Sirach in order to say about Jesus as the Word made flesh what the tradition had said long ago about Wisdom, thereby comparing, even equating, the two.[5]

Fifth, Wisdom's ultimate fate was to return to God, as the intertestamental Book of Enoch describes: "Wisdom could not find a place in

---

[5] Brown, ibid., draws a further parallel between John 3:13 ("No one has gone up to heaven except the one who has come down from heaven, the Son of Man.") and Wisdom-Solomon 9:16-17 and Baruch 3:29.

which she could dwell: but a place was found [for her] in the heavens. [So] Wisdom returned to her place and she settled permanently among the angels" (1 Enoch 42:1-2). This claim about Wisdom's fate provides the clue for the unique Johannine perspective on the death of Jesus as the "return to the Father" and the home going of the Word (John 14:2, 3, 12, 28; 16:5, 7, 17, 28; 17:11, 13). A sixth wisdom motif is the picture of Wisdom as the teacher, the revealer of truth that comes from the heavenly realm. Wisdom is a teacher of truth (WisSol 7:22; 9:16-18). She guides (WisSol 10:11), shows what pleases God (WisSol 9:10), gives life (Sir 4:11-12), and immortality (WisSol 6:18-20; Sir 4:13). Part of the uniqueness of her mode of teaching, especially important for John's Gospel, is her use of self-proclamations in the form of "I am" statements (Prov 8:3-36; Sirach 24). All these activities, and in particular the use of "I am" pronouncements, find counterparts in the portrait of Jesus drawn by the author of John.

Seventh, in the use of symbols the Fourth Gospel seems to prefer the very symbols that have prominence in the wisdom tradition. Bountiful food and drink (bread, water, and wine) are wisdom's gifts (Prov 9:2-5; Sir 24:19-21; Isa 55:1-3) as they were the special gifts of Jesus in John (John 2 , 4, 7, and 6). But the wine at Cana, the living water of the Samaritan well, the fountain welling up in the believer, and the bread and fish that Jesus multiplied for the crowd were no ordinary food and drink. They were signs of Jesus himself, given as food and drink leading to everlasting life.

An eighth wisdom motif concerns the manner in which Wisdom pursues her would-be followers. She takes the initiative, goes into the streets, and cries out *(krazo)* to by-standers to "Come and follow" (Prov 1:20-21; 8:1-4; WisSol 6:16). Again the Johannine Jesus adopts Wisdom's ways. He invites followers to "Come and see" (John 1:39), he seeks out potential believers (John 1:43; 5:14; 9:35), and cries out *(krazo)* to others in the streets and in the Temple precincts (John 7:28, 37; 12:44) as Wisdom did.

Ninth, just as Wisdom instructs disciples as her children (Prov 8:32-33) so Jesus instructs his disciples at the supper and addresses them as "My Children" (John 13:33), teaching them all they need to know to follow in his way.

A tenth motif shared by the wisdom tradition and the Gospel of John is the realization that Wisdom's offer of life can and would be rejected by some (Prov 1:24-25). The fate of those who reject Wisdom's gift is "doom, terror, distress" (Prov 1:26, 27), and, ultimately, death (Prov 1:32). It is no surprise then that the Johannine Jesus condemns the Jews who fail to believe in him with the harshest possible condemnation: "You belong to your father the devil . . ." (John 8:44). He is merely echoing Wisdom's condemnation of those who refuse to accept her offer of life.

In the same vein, the eleventh example of wisdom motifs in John is the belief that a confrontation with Wisdom provokes division between those who accept and follow after Wisdom's lead (Prov 8:17; WisSol 6:12; Sir 6:28) and those who reject her invitation (Prov 1:24-25). The gospel reminds us that the Johannine Jesus, like the figure of Wisdom, repeatedly provoked division among those who heard his message (John 7:43; 9:16; 10:19).

Finally, the last example of wisdom motifs in John is the parallel between the Spirit-Paraclete in John and the role that Wisdom plays as she enters into "holy souls and produces friends of God and prophets" (WisSol 7:27). John's emphasis on friendship, therefore, ("You are my friends if you do what I command you. I no longer call you slaves . . . I have called you friends" [John 15:14-15]) can be seen as just one more example of Wisdom's influence on the theological themes and portrait of Jesus in John.

Given these many shared motifs between the wisdom tradition and the Gospel of John, it is difficult to deny that for John Jesus was understood as the incarnation of Wisdom. If this claim is correct, then all that the tradition had affirmed of the figure of Wisdom—her oneness with God, her share in creation, her tent pitched among us as flesh, her salvific capacities and revelatory potential, her final return to God—all this could now be said of Jesus, the Word made flesh.

THE DIVINE "I AM"

Among these many examples of the ways in which wisdom motifs have shaped the story of Jesus in the Gospel of John, however, one feature is especially prominent and unique to the Fourth Gospel. Whereas in the Synoptic Gospels Jesus' characteristic mode of speech is the parable and the aphorism or short saying, in John Jesus speaks in the first person and makes numerous self-pronouncements throughout the Gospel. The predominance of these "I am" sayings, furthermore, is a key to John's unique Christology.

In the history of the interpretation of John scholars sought the origin of these "I am" sayings in the discourses of gnostic revealers, in the Mandean or Hermetic literature,[6] or especially in the self-proclamations of the Egyptian goddess Isis. It was thought that these Hellenistic parallels best explain the mode of Jesus' discourse in John. For example, a second-century inscription from Cyme in Asia Minor is often cited. It is a copy of an earlier Isis aretalogy (or series of claims on behalf of the goddess) from Memphis, and reads in part:

> I am Isis, the mistress of every land . . .
> I am eldest daughter of Kronos
> I am wife and sister of Osiris.
> I am she who findeth fruit for men . . .
> I am the Queen of rivers and winds and sea . . .
> I am the Queen of war
> I am the Queen of the thunderbolt . . .
> I am in the rays of the sun. . . .
> I am Lord [note masculine form] of rainstorms . . . etc.[7]

While there is similarity in form and function as "divine revelation" between these praises of Isis and Jesus' "I am" statements in John, the recent consensus sees the origin of Jesus' self-proclamations in John's Gospel to derive from elsewhere. The primary source for the "I am" pronouncements in John, it is now argued, must be found not in Hellenistic or pagan sources but in the speeches of YHWH in the First Testament (e.g., Exod 3:6; 3:14; 20:2; Isa 41:4, 13) and in the self-pronouncements of Personified Wisdom, exemplified especially in Proverbs 8 and Sirach 24.[8]

Three distinct groups of "I am" statements appear in the Gospel of John. The first are those sayings where "I AM" *(ego eimi)* occurs without a predicate—the absolute use of *I AM:*

> 8:24   For if you do not believe that I AM you will die in your sins
> 8:28   When you lift up the Son of Man, then you will realize that I AM
> 8:58   Before Abraham came to be I AM
> 13:19 . . . so that when it happens you may believe that I AM.

A second group of sayings are those where a predicate is understood though not expressed, such as:

> 6:20   He said to them, "It is I (lit. *ego eimi*). Do not be afraid."
> 18:5   [Jesus] said to them, I AM he (lit. *ego eimi*).

---

[6] See the summary in Rudolf Schnackenburg, *The Gospel according to St. John,* 2 vols. (New York: Crossroads, 1990) 2:81–83.

[7] The text in translation is found in Frederick C. Grant, ed., *Hellenistic Religions. The Age of Syncretism* (Indianapolis: Bobbs-Merrill Educational Publishing, 1953) 131–33.

[8] See C. K. Barrett, *The Gospel according to St. John,* 2nd ed. (Philadelphia: Westminster, 1978) 292–93; Brown, *The Gospel according to John,* 1-LX:533–38; Schnackenburg, *The Gospel according to St. John,* 2:81–86.

The third and largest group of "I am" sayings are those where Jesus speaks of himself in figurative or metaphorical language and where "I am" appears with a predicate nominative:

6:35, 51	I am the bread of life [living bread]
8:12; 9:5	I am the light of the world
10:7, 9	I am the gate [for the sheep]
10:11, 14	I am the good shepherd
11:25	I am the resurrection and the life
14:6	I am the way, and the truth, and the life
15:1, 5	I am the [true] vine.

For the earliest audience of the Fourth Gospel, steeped as they must have been in Israel's Scriptures, the words "I AM" could not but call to mind the words of Yhwh to Moses spoken again and again throughout the wilderness period, "I am Yhwh; I am he" (*ªni Yhwh; ªni hu).*[9] And when the Septuagint (LXX) translated these words of Yhwh they became in Greek, *ego eimi.* So, too, the words of Yhwh from the bush giving the enigmatic divine name became in Greek, *ego eimi.* There is little doubt that the author of John intended to give to Jesus the same self-identifying formula by which Yhwh chose to make the divine self known. The clearest proof of this claim in the gospel can be seen, amid full Johannine irony, in the scene of Jesus' arrest in the garden in John 18:1-6. The narrator tells us that Judas came with "a band of soldiers and guards from the chief priests and the Pharisees and went there with lanterns, torches and weapons" (John 18:3). The arresting party was armed and powerful, Jesus and his band of disciples, powerless and insignificant, by contrast. And yet, when Jesus steps forward and asks, "Whom are you looking for?" and they reply "Jesus the Nazorean," the narrator tells us that Jesus replied: "I AM [he]" whereupon they "fell to the ground" (18:5-6). At the literal level (reading "I am he" as a simple and straightforward response) this scene is absurd. It strains credibility to the breaking point and even borders on slapstick comedy reminiscent of the "Keystone Cops." But if we realize that the words of Jesus, *ego eimi* (I AM) communicate on the symbolic level and convey a divine theophany, then falling to the ground is the only appropriate response to this encounter.

The final group of "I am" sayings, those with predicate nominatives, come closest to Wisdom's way of self-proclamation and illustrate an

---

[9] See, e.g., Exodus 6:6, 7; 7:5; 20:1, 5, and the multiple uses of the same formula in Isaiah 41:4; 43:10, 25; 45:8, 18, 22; 46:4, 9; 47:8, 10; 48:12, 17; 51:12; 52:6.

even greater dependence on the wisdom tradition by the author of John. First, it is not insignificant that there are seven "I am" pronouncements by Jesus using, alternatively, the metaphors of bread (6:35, 48, 51), light (8:12; 9:5), sheepgate/door (10:7, 9), shepherd (10:11, 14), resurrection and life (11:25), way/truth/life (14:6), and finally the vine (15:1, 5). The number seven signals completeness and many would argue that the author has chosen this symbolic number not only for the "I am" sayings but also for the enumeration of the signs in the first half of the gospel.[10] This predilection for the symbolic number seven may well be another influence from the wisdom tradition which, in WisdomSolomon 7, uses the symbolic pattern of twenty-one (three groups of seven) to praise the attributes of wisdom, thus extolling her completeness and perfection. Like Wisdom, Jesus reveals the perfection of the one "from above" who sent him.

Second, in the list of these seven "I am" statements two are more abstract. As Schnackenburg noted, "Judged by the criterion of symbolism, 'the resurrection and the life' (11:25) and 'the way, the truth and the life' (14:6) have a more abstract and artificial ring, as though they had been developed out of the intellectual context specifically for Johannine theology."[11] We might, rather, have added other metaphors to the list such as Jesus as the spring of water (John 4 and 7) demonstrating both the explicit and the implicit claims of self-identity made by Jesus in the Gospel. Third, the multiplicity of the metaphors is also significant because it shows the universal character of the claims about Jesus in John. It is almost as though the Johannine Jesus says, "Whatever symbol best expresses your religious and human longings [water, bread, light, access door, shepherd, vine, etc.], I AM the fulfillment of your longings." Like the invitation of Wisdom extended to all in the streets and byways to "Come," so Jesus offers a similar invitation for all.

Rudolf Bultmann's classification of the "I am" sayings remains a helpful tool for understanding the function of these "I am" pronouncements. Bultmann differentiated four different types of statements: "(1) *Präsentationsformel,* or an introduction, answering the question, 'Who are you?' (2) *Qualifikationsformel,* or a description of the subject, answering the question, 'What are you?' (3) *Identifikationsformel,* where

---

[10] 1) Water into wine at Cana (John 2), 2) Cure of the Royal Official's son (John 4), 3) Cure of the paralytic (John 5), 4) Multiplication of the loaves (John 6), 5) Walking on the Water (John 6), 6) Cure of the Blind Man (John 9), and 7) Raising of Lazarus (John 11).

[11] Schnackenburg, *The Gospel According to St. John,* 2:80.

the speaker identifies him/[her]self with another person or thing, and
(4) *Rekognitionsformel,* or a formula that separates the subject from
others. It answers the question, 'Who is the one who . . . ?'"[12] Using
Bultmann's categories, the latter two, the "Identification formula" and
the "Recognition formula" are both present in Jesus' "I am" statements
in John. Together these two categories explain how Jesus reveals something of the mystery of his being and shared life with God and how he
mediates that life to others under the symbols of "bread, light, shepherd, vine, etc."

Not only Jesus' mode of speaking in the first person, but also the
content of many of the metaphors themselves betray the influence of
the wisdom tradition. When Jesus declares "I am the bread of life"
(John 6:35) and contrasts the nourishment he gives with the manna in
the wilderness he echoes numerous related biblical passages about
God feeding the Israelites: Exodus 16:4, 15; Psalm 78:24; Nehemiah
9:15; WisdomSolomon 16:20, to name just a few. But, like her Creator,
Wisdom, too, sets her table and invites all to enter and eat: "Wisdom
has built her house, . . . she has dressed her meat, mixed her wine . . .
she has spread her table. Come, eat of my bread, and drink of the wine
I have mixed!" (Prov 9:1, 5). In the wisdom tradition of Israel, moreover,
the manna was interpreted as word and instruction, and so John's
Jesus is the incarnate Word of God and, hence, the real food. Moreover,
Jesus' bread of life discourse trumps Wisdom's claim that "[the one]
who eats of me will hunger still, [the one] who drinks of me will thirst
for more" (Sir 24:20) when Jesus proclaims: "whoever comes to me will
never hunger, and whoever believes in me will never thirst" (John 6:35;
cf. 4:14).

Likewise, the metaphor of light recalls not only the depiction of Wisdom herself as the reflection of God's light (WisSol 7:26) but also Wisdom's equating of the Law with the light to enlighten the path of human
holiness (WisSol 18:1-3). When Jesus proclaims "I am the light of the
world," therefore, he both reveals his true identity as Wisdom incarnate
and invites all who seek enlightenment to turn to him, the only true
light. In a final self-proclamation Jesus also identifies himself as the
"true vine." The image of the vine/vineyard is ubiquitous in Palestine
and served as a ready image to speak of Israel as "God's vineyard" (Isa
1:8; 3:14; 5:1, 3, 4, 5, 7,10; 27:2; Jer 12:10). Wisdom, too, proclaims her

---

[12] See Brown, *The Gospel According to John,* 1:533; Rudolf Bultmann, *The Gospel of
John: A Commentary* (Oxford: Blackwell, 1971) 167.

identity as one who has "taken root" among God's people. "I am like a vine putting out graceful shoots, my blossoms bear the fruit of glory and wealth. Approach me, you who desire me and take your fill of my fruits" (Sir 24:17-18 NJB). Jesus, therefore, echoes again the voice of Wisdom when he says, "I am the vine . . . whoever remains in me and I in [them] will bear much fruit" (John 15:5).

WISDOM AS THE KEY TO JOHANNINE CHRISTOLOGY

John's Gospel contains the highest christological affirmations about Jésus in the Second Testament. From the prologue's praise of the pre-existent one who "pitched a tent" with us and "became flesh" as the one sent from God, through all the encounters of the Gospel where Jesus invites others to "Come and see," we see Jesus as the great revealer of the glory and the things of God. He is the ladder that connects earth with heaven (John 1:51), the perfect mediator of God's life to others. Like Wisdom, both his origin and his destiny are with God. Like Wisdom, his invitation is an invitation to life—eternal life, that is, life with God. He shows the way to God because, like Wisdom, he has come from God and was with God in the beginning. In rich, abundant metaphors of food and drink, like Wisdom, Jesus offers to feed all who come and stay where he dwells. Like Wisdom, he speaks with the authority of God and does what God does. Just as the affirmations of the "divine" qualities of Woman Wisdom (WisSol 7:22-30) strained the theological categories of Jewish faith in its day, so John's portrait could claim for Jesus a oneness with God ("The Father and I are one," 10:30) that would become the raw data of christological reflection for the next four centuries up to the Council of Chalcedon in 451 C.E. Finally, that Jesus' identity and mission were interpreted by John's Gospel through the lens of the personified figure of Woman Wisdom should remind us that God's image in human form is, as the book of Genesis affirmed long ago (Gen 1:27), always inclusive of male and female. All this is the truth and the wisdom of the Johannine Jesus, and that truth, the Gospel tells us, will indeed make us free.

5

Creation as a
Divine-Human Collaboration

HERMAN E. SCHAALMAN

Few would quarrel seriously with the assertion that the opening chapters of the Bible deal with the mystery of existence in a fascinating and often profound manner. The unfolding tale of how the first something came into being, and how in stately procession other major phenomena of reality appeared by no other means than the Divine Word, has maintained its capacity to attract and hold the imagination, and often the belief, of untold millions of readers or hearers of the *maʾase bereshit,* the Story of Creation.

In particular, the progressive sequence from light to earth and, then, to water and land with its uncountable forms of life, climaxing in the creation of Adam, has arrested our attention and elicited astonishment and wonder. And when the text then records that after the arrival of Adam, God judged the work as *tov mʾod,* "very good" (Gen 1:31), an unmistakable escalation beyond earlier statements of approval which were only *tov,* "good" (Gen 1:12), the intent of the story becomes unavoidably clear. Adam, we human beings, are the "crown" of creation, God's masterpiece.

Not totally surprising, the text then continues to assert: "The heavens and the earth were finished, and all their array" (Gen 2:1f.). And further: ". . . and He ceased on the seventh day from all His work which He had done" (Gen 2:2). *"Vayishbot,"* i.e., God made *Shabbat.* No matter how manifold, really infinite the results of creation had been prior to Adam, it is only when the human appears, the text holds, God is finished with "working." Only then does God enjoy His *Shabbat.* Does God "deserve" a *Shabbat?* Apparently; for no matter how complicated and multifarious all previous creative acts may have been calling for the creative genius of the deity, the enjoyment of *Shabbat,* of rest, of completing is withheld until after Adam's appearance. To put it differently, God does not "enjoy" or "know" of *Shabbat* until you and I are present. Does God's *Shabbat* then depend on us? Why not! God does not appear to be totally, radically self-sufficient. God is inherently, unavoidably linked to us, beginning with the divine rest of *Shabbat.*

This is, of course, an unavoidable nexus, since all of our thoughts, statements, texts are naturally in the form of words, language, human language. We only "know" God, express God in human terms, in the only manner which is possible for us, i.e., as human. "God" in itself is inaccessible, unknowable, inexpressible to us. God is ineffable. Hence all our words about God are our words, at best approximations, hints, inherently imperfect articulation of what is by our understanding essentially beyond us, beyond our humanness to grasp in totality, completely. This basic recognition makes it incontrovertible that God depends on us to be known, sought, probed, rejected, loved. So the Talmud asserts repeatedly: *"dibra torah bilshon bᵊne adam;* Torah speaks in human language."

Standard translations of the Hebrew original distort the text when they read: "And God blessed the seventh day and declared it holy, because on it God ceased from all the work of creation which He had done" (Gen 2:3). A correct rendition of the last phrase would read: ". . . for on it He rested *(shavat)* from all His work which God had created to do." Creation was not finished, not complete. God ceased, but only "to do!"

And to whom would the task be assigned "to do" if not to that final "work" which crowned the previous creative act, Adam. God could take a *Shabbat* because now Adam, you and I were available "to do," to carry on, to work toward completion, toward finishing, toward *tikkun olam* the repairing, the fixing of the world, for creation was left imperfect; intentionally, unavoidably?

Could God have enjoyed this first, prototypical *Shabbat* without Adam? Not the way the text is structured. God did not "rest" after creating the sun and moon, or the sky and earth, etc., etc. Only after Adam, prototype of all humans, is created *vayishbot* God "rested," God created *Shabbat.* Right from the very beginning of this most remarkable story of the coming into being of all reality the biblical text suggests, or at least allows the understanding that between God and Adam there exists a relationship of connectedness, of mutuality, of reciprocity. It is the model of what will much later be expressed as the *berit,* the Covenant, for the Covenant is also, uniquely, inescapably a relation of mutuality, of interdependence, and not only of humans on God, but covenantally of God also on humans. For, what would happen to God if there were no humans? Who would pray, search, love, argue, seek to understand?

Lest such a reading of the Creation Story is deemed to be too tendentious, too exquisite, or simply wrong, let us notice another portion of this foundational tale which, curiously, had little or really no place in the commentaries: "The Lord God said, 'It is not good for man to be alone; I will make a fitting helper for him.' And the Lord God formed out of the earth all the wild beasts and all the birds of the sky, and brought them to the man to see what he would call them; and whatever the man called each living creature, that would be its name. And the man gave names to all the cattle and to the birds of the sky and to all the wild beasts; but for Adam, no fitting helper was found" (Gen 2:18f.). Quite apart from its charm, this is a most astounding pericope. We are well beyond the verse which describes the original creation of Adam and the subsequent divine attestation that this event was "very good" to learn only now that "it was not good for man to be alone." God did not anticipate this defect? God did not know that Adam needed a companion, a human other? Does the text tell us that God is "learning" something new?

Apparently, and midrashic commentaries and elaborations insist that God had the animals parade before Adam in pairs, male and female. Was only Adam to remain single, nonsexual, while all along the animals were clearly paired, though in Eden there would have been no procreation anyway? Did God only "discover" then when the paired animals passed by Adam that he had not been favored similarly, so that "it was not good for man to be alone?" No omniscience here.

There is more: "And the Lord God formed . . . and brought them to the man to see what he would call them" (Gen 2:19). They had no "names" 'til then, no definition, no identity. They were "wild beasts . . . birds of the sky" undifferentiated, existing helter-skelter in unnamed mass. God, so the text tells us, "brought them to the man to see what he would call them." Taxonomy is a human creation. Differentiation comes from Adam. Names, identities, definitions stem from the human collaborator who functions in this fashion at the behest of God who did no such thing beforehand. Or, perhaps such naming is reserved only for us humans, only possible for us, an inescapable consequence of being human. Who else would call a giraffe a giraffe?!

It is human insight, discrimination, language that identifies and, therefore, also orders the hitherto chaotic lack of differentiation into cosmos. The structure of the nonhuman biomass, while created by God, becomes only manageable, orderly, structured by the only-then-human, Adam. Or to put it differently, the created "order" does not emanate from God. It derives from Adam. From the very first moment of reality, as depicted in the biblical record, the human is involved radically, indispensably. Even the most obvious and necessary consequences of the divine initiative invites, nay needs human assistance, involvement, collaboration. There would be no "world" without human participation, without human language to designate and, hence, render it intelligible. The world, reality, as it is, is not only God's "work." It is the result of a cooperative process, of human participation in a fundamental, radical fashion. God only "rested." There was more to be done. And God did not do it alone. God did it with us; does it with us. We were then, as we are now, indispensable. Even the mystery of Creation is not God's solely. We are involved and needed nearly from the very beginning.

When read and interpreted this way, the biblical text portrays an image of God fundamentally at odds with much later constructions formulated in absolute terms by way of Greek philosophic influences. This view of God is radically relative, interactive with the human. It is not of the absolute monarch, or the equally foreign First Cause, or the construct of omnipotence and omniscience. It is the opposite. It is God becoming, suffused in the unfolding fabric of reality, of existence, of life. It is God with us, *immanuel*.

This is proof, if such is still necessary (and perhaps it is), that the Bible is not a book of philosophy, of one theology or belief, but from the opening chapters and onward the story of how humans learn of God who may not be encased in formulae devised by humans, but whose presence floods reality, the omnipresent God.

To top the story off, when Adam can find "no fitting helper" (Gen 2:21f.), then "The Lord God cast a deep sleep upon the man . . . and He took one of his ribs . . . and . . . fashioned into a woman the rib . . . ." God "learns" that Adam needs his like as a companion. And of this Adam could say "This one at last is bone of my bones and flesh of my flesh" (Gen 2:23). The discovery of woman, her naming such, hence her identity is Adam's contribution to God's creation. Again it is Adam who "names."

The fundamental verity of being human is that it is a collaborative event between God and us.

6

Creation, Revelation, and Redemption

Recovering the Biblical Tradition as a Conversation Partner to Ecology

CAROL J. DEMPSEY, o.p.

INTRODUCTION

Many years ago, Meister Eckhart wrote:

> Apprehend God in all things,
>   for God is in all things.
> Every single creature is full of God
>   And is a book about God.
> Every creature is a word of God.
> If I spent enough time with the tiniest creature—
>   even a caterpillar—
> I would never have to prepare a sermon.
> So full of God is every creature.[1]

---

[1] This quotation has been translated by Matthew Fox, who renders it as a poem. See Matthew Fox, *Meditations with Meister Eckhart* (Santa Fe: Bear & Company, Inc., 1982) 14.

The wonder that led Eckhart to see God in everything that exists, even in the most insignificant of creatures is, I suspect, the same wonder that led the ancient psalmist to proclaim:

> The heavens are telling the glory of God;
>   and the firmament proclaims God's handiwork.
> Day to day pours forth speech,
>   and night to night declares knowledge.
> There is no speech, nor are there words;
>   their voice is not heard;
> yet their voice goes out
>   through all the earth,
> and their words to the end of the world (Ps 19:1-4).[2]

It is this same sense of awe, this same sense of marvel about creation that has inspired poets, prophets, and preachers down through the ages. It is this same sense of wonder and awe that needs to be recovered today if we are ever going to break the downward spiral of violence and destruction that is raping our planet, anesthetizing hope, and snuffing out the very spirit of God that pulsates at the center and heart of all that is.

Theologian Elizabeth Johnson lays out the picture clearly:

> The capacity of the planet to carry life is being exhausted by . . . human habits. Not only is our species gobbling up resources faster than earth's ability to replenish itself, but our practices are causing damage to the very systems that sustain life itself; holes in the ozone layer, polluted air and rain, clear-cut forests, drained wetlands, denuded soils, fouled rivers and lakes, polluted patches of ocean. Appallingly, this widespread destruction of habitats has as its flip side the death of creatures that thrive in these ecosystems. By a conservative estimate, in the last quarter of the 20th century, 20 percent of all living species have become extinct. When these creatures, these magnificent plants and animals, large and small go extinct, they never come back again. We are killing birth itself, wiping out the future of fellow creatures that took millions of years to evolve. We live in a time of great dying off caused by human hands.
>
> On the one hand, we gaze in wonder at the world; on the other hand, we are wasting the world. This is a sign of our times and should be seriously reckoned with by people of faith. But, the odd thing is that, with some notable exceptions, many religious people and church business as a whole go on ignoring the plight of the earth.[3]

---

[2] Biblical texts used in this article are from the New Revised Standard Version of the Bible.

[3] Elizabeth Johnson, "God's Beloved Creation," *America* (April 2001) 9.

Some may be weary of hearing about saving the rainforest, drudging the Hudson, cleaning up the Columbia, and protecting the salmon. Yet, the reality is that we are being engulfed by an ecological crisis, but a crisis that positions preachers, teachers and scholars on the threshold of opportunity for reclaiming our rich biblical and theological heritage. In particular, the current ecological crisis positions us to revision the anthropocentric lenses employed for considering creation, revelation and redemption as we hear anew the biblical text as a conversation partner to theology.

This article attempts to model this conversation between the Bible and theology. Before entering into conversation, however, we have to consider our common social location—planet earth—and engage in critical theological reflection of the biblical text from that vantage point as we ponder the questions of revelation and redemption. Only then will we consider why we need to be concerned about the devastation of creation, and only then will we be able to develop a new environmental ethic whose driving force will be deeper than sustainability. We need to touch the oneness of all and to let Isaiah's vision (Isa 65:17-25) of a new heavens and a new earth move us to praxis. While pondering, writing, and reflecting are necessary, these are not enough. The biblical text calls us to be the poets, prophets, and storytellers of a new day. The words on the page invite us to roll up our sleeves, put forth a vision, and walk in the light of that vision. And where do we begin? We begin at the beginning—with creation.

CREATION: GENESIS 1–2

The first poetic narrative of the Bible, the creation story in Genesis 1–2, makes clear several points. First, according to the ancient people and biblical writers, creation was intrinsically good and affirmed by God. Repeatedly we hear, "And God saw that it was good (1:10, 12, 18, 21, 25). The culminating statement comes in Genesis 1:31: "God saw everything that he had made, and indeed, it was very good." Sea creatures and birds (1:21), human beings (1:28), and the seventh day (2:3) are also the recipients of divine blessing. The sea creatures, birds, and human beings have a common task associated with God's blessing—they are to be fruitful and multiply (1:22, 28).

A second clear point of this biblical narrative is that plants, animals, and the first human being have a common divine origin in God and a common natural origin in the "ground" (2:7, 9, 19). God's *ruach* ("wind," "breath," "spirit"), present at the beginning of creation (1:1), brings everything to life (1:30; 2:7; cf. Ps 104:29-30).

Third, the six consecutive days of creation in the biblical texts present creation as a series of interdependent relationships. With respect to the consecutive days of creation in the P narrative (Gen 1:1–2:4a), it is important to consider how the ancient people understood time—not in terms of linear calendars as we do today, but rather, in terms of seasons and cycles and sundials. The sixth day, then, becomes part of the ordered cycle of creation, and human beings created on the sixth day can then be seen as part of the vision and cycle of creation. Human beings become part of the whole picture of creation and not the main focal point of it. This rereading allows the sense of interdependence and relationship that exists among all dimensions of creation to emerge from the biblical text.

Together, the P narrative and the J narrative (Gen 2:4b-25) present us with a tapestry of interconnected relationships and a sense of unity that exists between that which is divine, human, and nonhuman. The "light" and "darkness," separated one from another, function as "day" and "night," respectively, and thus complement each other (1:3-5). The dome in the middle of the waters separates one body of water from another (1:6-8). The water under the sky is gathered into a single basin—the sea—which allows for the dry land to appear (1:9). The sea teems with and becomes "home" to the sea creatures as the birds of the air "fly above the earth across the dome of the sky" (1:20-21). The earth brings forth vegetation (1:9) which becomes food for human beings and the rest of earth's creatures (1:29-30). Lights—the sun, the moon, and the stars—appear in the dome of the sky and separate day from night (1:14-19). The fields and grass need the rain to water the earth before they can sprout up (2:4-5). A river that flows out of Eden waters the garden that God had planted in the midst of creation (2:8, 10). Animals are intended to be helpers as partners for the human being (2:18-19), a divine intention more closely realized through the creation of and relationship between the two human beings (2:20-23). In the P narrative human beings are created together with gender differentiation (1:27); in the J narrative (2:4b-25) the second human being is created from the first one's rib (2:22). Bone from bone and flesh from flesh (2:22), they become one flesh (2:24). Both are naked and unashamed (2:25).

Fourth, the human being is given a charge. Placed in the garden by God, the human being has the task of tilling and keeping the garden (2:15). In "Restoring Creation for Ecology and Justice," *A Report*

Adapted by the 202nd General Assembly of Presbyterian Churches (U.S.A.), "tilling" and "keeping the earth" are interpreted through an ecological lens:

> "Tilling" symbolizes everything we humans do to draw sustenance from nature. It requires individuals to form communities of cooperation and to establish systematic arrangements (economies) for satisfying their needs. Tilling includes not only agriculture but mining and manufacturing and exchanging, all of which depend necessarily on taking and using the stuff of God's creation. . . . "Keeping" the creation means tilling with care–maintaining the capacity of the creation to provide sustenance for which tilling is done. This, we have come to understand, means making sure that the world of nature may flourish, with all its intricate, interacting systems upon which life depends.[4]

The care for the garden mentioned in the J narrative complements another task given to the human beings in the earlier P narrative: to have "dominion over the fish of the sea, and over the birds of the air, and over the cattle, and over all the wild animals of the earth, and over every creeping thing that creeps upon the earth" (1:26) and to "subdue" the earth. Bruce Birch, Walter Brueggemann, Terrence Fretheim, and David Petersen argue that

> the command to have dominion (1:28), in which God delegates responsibility for the nonhuman creation in power-sharing relationship with humans, must be understood in terms of care-giving, not exploitation (see *radah* in Ps 72:8-14; Ezek 34:1-4). The verb *subdue*, while capable of more negative senses, here has reference to the earth and its cultivation and, more generally, to the becoming of a world that is a dynamic, not a static reality.[5]

This brings us now to the point of "stewardship." Elizabeth Johnson acknowledges the strengths and weaknesses of the stewardship model.

---

[4] This statement is taken from "Restoring Creation for Ecology and Justice," *A Report Adopted by the 202nd General Assembly* (1990), Presbyterian Church (U.S.A.) 7. Phyllis Trible's comments on "to till" and "to keep" are most insightful. She notes that ". . . to till and to keep, give the earth creature power over the place in which Yahweh God put it." *God and the Rhetoric of Sexuality; Overtures to Biblical Theology* (Philadelphia: Fortress Press, 1978) 85. Trible highlights the connection between humanity and botany. She also makes the point that "to till the garden is to serve the garden; to exercise power over it is to reverence it" (ibid.). Finally, she argues that "to till and to keep, connote not plunder and rape but care and attention" (ibid.).

[5] Bruce C. Birch, Walter Brueggemann, Terrence Fretheim, and David L. Petersen, *A Theological Introduction to the Old Testament* (Nashville: Abingdon Press, 1999) 50.

She recognizes that the model "keeps the structure of hierarchical dualism but calls for human beings to be responsible caretakers or guardians of the earth and all its creatures."[6] Johnson also notes that in this model, "humanity is still at the top of the pyramid of being but has a duty to protect and preserve what seems weaker and more vulnerable."[7] Johnson sees this model as an "improvement over the absolute ruler/kingdom-kingship model," for it "guarantees a modicum of respectful use of the earth."[8]

This theme of care is picked up by the Wisdom tradition in Psalm 104 where the psalmist catalogs God's care for all creation while proclaiming

> O Lord, how manifold are your works!
> In wisdom you have made them all;
> The earth is full of your creatures (v. 24).

In summary, Genesis 1–2 speaks of (1) the intrinsic goodness of creation, (2) the relationships that exist among all aspects of creation in general, (3) the relationships that exist between God, human beings, nonhuman beings, and the natural world in particular, and (4) the care of creation.

CREATION AND THE WISDOM TRADITION

The book of Sirach describes wisdom as originating and dwelling with God (1:1, 9; 24:9). Wisdom was fashioned before the ages, in the beginning, before all things were created, and will exist for all ages (1:4;

---

[6] Elizabeth Johnson, *Women, Earth, and Creator Spirit* (Mahwah, N.J.: Paulist, 1993) 30.

[7] Ibid.

[8] Ibid. I agree with Johnson's assessment of this model and also agree with her idea that the stewardship model "misses the crucial aspect of human interdependence upon that which we steward" (ibid., 31). I suggest that a new model is needed, along with a new ethic, one that acknowledges and respects the intrinsic goodness and interdependence of all creation, whereby all creation is valued and cherished not because the biblical text calls us to be stewards, but rather, first and foremost because creation is, in and of itself, good.

With Johnson, I agree that we need to get to the point of realizing that "the natural world has given birth to all living things, and sustains us all. It is the matrix of our origin, growth, and fulfillment. Articulated within a religious perspective, the kinship stance knows that we humans are interrelated parts and products of a world that is continually being made and nurtured by the Creator Spirit. Its attitude is one of respect for the earth and all living creatures including ourselves as a manifestation of the Spirit's creative energy; its actions cooperate with the Spirit in helping it flourish. What goes on in this stance is neither a sentimental love of nature nor an ignorance that levels all distinctions between human beings and other forms of life. Rather what is involved is a recognition of the truth: human existence is in fact one with the immensity of all that is" (ibid., 31).

24:9; cf. Prov 9:22-31). God poured out Wisdom upon all God's works, upon all the living, and upon those whose love is for God (1:10). In Sirach 16:24–17:24, the biblical writer praises God's wisdom found in the order of creation. In the Wisdom of Solomon, wisdom is a "breath of the power of God," a "pure emanation of the glory of the Almighty" (7:25), a "reflection of eternal light," a "spotless mirror of the working of God," and "an image of [God's] goodness" (v. 26). Proverbs gives voice to Wisdom and declares that "she"[9] was God's first act of creation and preceded the creation of the heavens and the earth (8:22-26), that she was beside God at the time of creation (8:27-30; cf. 3:19-30), and that she was God's delight daily, rejoiced before God, rejoiced in God's inhabited world, and delighted in the human race (8:30-31). The book of Job declares that creatures and plants have the potential for teaching human beings something about the majesty and wonder of God:

> But ask the animals, and they will teach you,
>> the birds of the air, and they will tell you;
> Ask the plants of the earth, and they will teach you,
>> and the fish of the sea will declare to you.
> Who among all these does not know
>> that the hand of Yʜwʜ has done this?
> In his hand is the life of every living thing
>> and the breath of every human being (12:7-10).

All of creation, then, is made with Wisdom, is full of Wisdom, and has the potential to impart Wisdom. But how does God view all that has been made? The writer of the book of Wisdom gives us an insight:

> For you so love all things that exist,
> and detest none of the things that you have made,
> for you would not have made anything if you had hated it.
> How would anything have endured if you had not willed it?
> Or how would anything not called forth by you have been preserved?
> You spare all things, for they are yours, O Lord, you who love the living.
> For your immortal spirit is in all things (11:24–12:1).

All of creation makes manifest the Wisdom and glory of God and is loved by God who has compassion for every living thing (Sir 18:13).

Reflecting upon the role of God's Spirit in creation, theologian Walter Kasper notes:

> Since the Spirit is divine love in Person, he is first of all, the source of creation, for creation is the overflow of God's love and a participation in

---

[9] Throughout the wisdom tradition and writings, "Wisdom" is personified as a woman.

God's being. The Holy Spirit is the internal (in God) presupposition of communicability of God outside of himself. But the Spirit is also the source of movement and life in the created world. Whenever something new arises, whenever life is awakened and reality reaches ecstatically beyond itself, in all seeking and striving, in every ferment and birth, and even more in the beauty of creation, something of the being and activity of God's Spirit is manifested.[10]

Denis Edwards picks up on Kasper's point and suggests that,

The Creator Spirit's role is to enable each creature to be and to become, bringing each into relationship with other creatures in both local and global ecological systems, and in this process of ongoing creation, relating each creature in communion within the life of divine Persons-in communion.

This means that forests, rivers, insects, and birds exist and have value in their own right. They are not simply there for human use. They have their own integrity. They exist as an interdependent network of relationships in which each creature is sustained and held by triune love. They manifest the presence of the Spirit as the ecstasy and fecundity of divine love. . . .

The presence of the Spirit cannot be limited by human expectations. . . . The Spirit transcends our humanity and its preoccupations with itself, embracing all creatures. . . . The Spirit transcends our humanity and its preoccupations with itself, embracing God's creatures. . . . The Spirit pervades the whole universe and sees to "the depths of God" (1 Cor 2:10). To be in communion with this Spirit is to be in communion with the whole of creation. . . . All this is suggesting that Earth reveals. It is the place of encounter with the Holy Spirit.[11]

Thus all creation can become for us a gift that has the potential of revealing to us something about the Divine.

REVELATION IN LIGHT OF CREATION AND
THE WISDOM TRADITION: A THEOLOGICAL PERSPECTIVE

Having considered creation in the biblical tradition, it is now time to sample what theologians have to say about the revelation of God in light of creation. For Irenaeus of Lyon creation is the first stage of revelation: "Through creation itself, the Word reveals God the Creator;

---

[10] Walter Kasper, *The God of Jesus Christ* (London: SCM, 1983) 227.

[11] Denis Edwards, ed., "For Your Immortal Spirit Is in All Things," *Earth Revealing, Earth Healing* (Collegeville: The Liturgical Press, 2001) 64–65.

through the world, it reveals the Creator as builder of the world; through its work, it reveals the Workman; through the Son, it reveals the Father."[12] Thomas Aquinas makes it clear that human reason can come to understand something of the nature of God from creation.[13] Edward Schillebeeckx comments on this point:

> What God has created is not merely a neutral nature which [a person] can analyse by means of natural science, but created nature, i.e., the expression of a divine plan which extends beyond "nature." In that case, this nature cannot, theologically speaking, be separated from the divine plan. . . . this nature is also the living expression of a divine decision which in the last resort proves to be a will for salvation. Creation and consummation are embraced by one great divine plan.[14]

John Haught adds this insight to the conversation:

> Revelation is the ongoing outpouring of God's creative, formative love into the whole world. In this sense it has a "general" character, and it is even constitutive of all things. Thus the idea of revelation in contemporary theology tends to converge with the biblical theme of creation. Creation itself is already the self-revelation of God.[15] . . . From the moment of its creation the universe has "felt" the outpouring of God's own being into itself, arousing it to reach out toward ever more intense modes of fulfillment. This divine self-donation is known as "universal" or "general" revelation.[16]

CREATION, REDEMPTION, AND THE PROPHETIC TRADITION

Creation is a main theme not only in the Wisdom tradition but also in the prophetic tradition. Prophetic literature is concerned in particular with the violation of covenant and the consequences that follow. When covenant is broken, human beings and the natural world suffer. One example of this connection can be found in Isaiah 24:4-6:

[12] Irenaeus, *Adversus Haereses* IV, 6,6; also II, 6,1; 27, 2; III, 25, 1; V, 18, 3 as cited in Rene Latourelle, "Irenaeus," in *Theology of Revelation* (Staten Island, N.Y.: Alba House, 1966) 99–100.

[13] *Summa Theologica,* Part One, questions 1–13.

[14] Edward Schillebeeckx, *Christ: The Experience of Jesus as Lord,* John Bowden, trans. (New York: Crossroad, 1980) 516.

[15] John Haught, "Revelation," *The New Dictionary of Theology,* Joseph A. Komonchak, Mary Collins, and Dermot A. Lane, eds. (Collegeville: The Liturgical Press, 1987) 884.

[16] Ibid., 887.

> The earth mourns and fades,
>> the world languishes and fades;
>> both heaven and earth languish.
> The earth is polluted because of its inhabitants,
>> who have transgressed laws, violated statutes,
>> broken the ancient covenant.
> Therefore a curse devours the earth,
>> and its inhabitants pay for their guilt;
> Therefore they who dwell on earth turn pale,
>> and few [people] are left.

This text suggests that there is a relationship between human sinfulness and the suffering of the earth or the natural world. As we reflect on Isaiah's words, we cannot help but be struck by the fact that although these words are ancient and metaphorical—referring to idolatry and a military invasion—it is as if they were written today. Is not our earth polluted because of its inhabitants? Have not some of us transgressed laws, violated statutes, broken the ancient covenant? Isn't the curse of eco-injustice devouring our earth? Have some members of the human community become so numb that they no longer have any ethical conscience or sense? Aren't many of us groping and turning pale from loss of spirit and passion—a loss that takes the form of a paralyzing malaise? Will the human race be few people left as it heads toward self-inflicted extinction, now foreshadowed by the extinction of so many wonderful plant and animal species? Is there hope for all of us?

The prophets show us through their proclamations that there is an inherent link between creation and redemption, specifically, that the redemption of humankind is connected to the restoration of the natural world through divine promise.[17] One text that exemplifies this link is Isaiah 44:1-5. In this text we hear God, speaking through the prophet, offering reassurance to sinful Israel (vv. 1-2) and revealing a redemptive plan of action on their behalf (vv. 3-5). Israel's redemption from sin is connected to the cultivation and renewal of the natural world. We hear a similar theme in Ezekiel 36:33-35 where the prophet declares that God will not only cleanse the people from their iniquities but also transform their desolate land. These two passages are only two of many within prophetic literature that develop these themes and link redemption and restoration.[18]

---

[17] I understand the term "redemption" as synonymous to "liberation" and "restoration."

[18] For a comprehensive study on cosmic redemption, see Carol J. Dempsey, "Hope Amidst Crisis: A Prophetic Vision of Cosmic Redemption," *All Creation Is Groaning: An In-*

REDEMPTION IN LIGHT OF CREATION:
A THEOLOGICAL PERSPECTIVE

For contemporary theological reflection upon creation and redemption, we turn to Francis Schüssler Fiorenza. He offers this reflection:

> Redemption is a central category of Christian theology, for it explicates the Christian proclamation of Jesus as Christ, as our Redeemer and Savior. The English word 'redemption' literally means a buying back. . . . The term "redemption" is best understood as a liberation from one state to another: from bondage to liberation. Redemption is the act or process by which the change takes the place.[19]

He goes on to note that "Redemption therefore does not simply blot out the punishment of sin, but it frees and liberates humans from the cosmic power of law, sin and death. Although present, redemption is also future and it embraces not only humans, but the whole creation."[20]

Key here is Fiorenza's emphasis that redemption embraces "the whole creation." If we understand this to be true, then is redemption only limited to the here and now, or do we understand redemption as all creation going to God? If the latter is true, then do we reconfigure our theological beliefs and teachings and way of life to reflect this truth?

CONTEMPORARY HERMENEUTICAL IMPLICATIONS
FOR THE ECOLOGICAL QUESTION AND CRISIS TODAY

As we have heard from various strands of the biblical and theological traditions, creation is the starting point for revelation. Furthermore, redemption embraces not only human beings but also all creation. This suggests the possibility of considering the concepts of revelation and redemption anew, not in the traditional anthropocentric sense but in the cosmological sense. In that regard, Elizabeth Johnson offers the following comment:

> The great act of redemption through the death and resurrection of Jesus Christ is intended not just for humanity but for the whole cosmos, for

---

terdisciplinary Vision for Life in a Sacred Universe, Carol J. Dempsey and Russell A. Butkus, eds. (Collegeville: The Liturgical Press, 1999) 269–84; Dempsey, "Cosmic Redemption: Embracing the Prophetic Vision and Spirit," *Hope Amid the Ruins: The Ethics of Israel's Prophets* (St. Louis: Chalice Press, 2000) 119–25.

[19] Francis Schüssler Fiorenza, "Redemption," *The New Dictionary of Theology,* 836.

[20] Ibid., 840.

God reconciled "all things, whether on earth or in the heavens, making peace through the blood of his cross" (Col 1:20). The view that the earth bears religious importance is also rooted in the rich biblical themes of incarnation (the Word becomes flesh, and so the divine joins with the matter of this world), Eucharist (sharing through bread and wine in the body and blood of Christ) and hope centered in Christ, that in the future the cosmos will enjoy fullness of new life in the glory of God. In view of this faith, Christians must inevitably conclude that "the ecological crisis is a moral problem."[21]

It is the creation story, then, that sets the agenda for a way of life that shifts the paradigm from salvation history conceived in an anthropocentric sense to one conceived in a cosmic sense. Claus Westermann, Helmut Schmid, Rolf Knerim, and James Barr, among others, have signaled the emergence of a new paradigm of biblical interpretation that accepts creation as the primary horizon of biblical theology. With this new paradigm, the current ecological crisis becomes evermore pressing and serious, for what we are experiencing is not only the loss of countless species but also something of the Spirit and presence of God as Being who dwells at the heart of and in the midst of all creation. From this perspective, it is both reasonable and disturbing to suggest that the central question for the twenty-first century is the God-question, and the central ethical issue is an ecological one.

Thus it is good for us to work and pray in the spirit of this ancient Jewish prayer:

And God saw everything that He had made,
    And found it very good.
And he said: This is the beautiful world
    That I have given you.
Take good care of it; do not ruin it.
It is said: Before the world was created,
    The Holy One kept creating worlds and destroying them.
Finally he created this one, and was satisfied.
He said to Adam [and Eve]: This is the last world I shall make.
I place it in your hands: hold it in trust.[22]

---

[21] Johnson, "God's Beloved Creation," 9.

[22] Cited in *Earth Prayers,* Elizabeth Roberts and Elias Amidon, eds. (San Francisco: HarperSanFrancisco, 1991) 62.

7

Three Who Loved Wisdom

AGNES CUNNINGHAM, S.S.C.M.

INTRODUCTION

What is philosophy, if not the love of wisdom? The ancient Greeks bequeathed to us the legacy of a word and a concept that reflected an attempt on the part of that ancient people to discover the ultimate cause or causes of the universe and to understand who and what the human being was in the cosmos. Even after Rome was recognized as the political ruler of the Mediterranean world, the culture of Greece continued to shape the ethos of the Empire. Greece, though conquered, was a conqueror. Greek became the universal language for communication as the classical *polis,* the city-state, was replaced in the Hellenistic world by the "new city," following the model founded by Alexander (d. 323 B.C.E.) in Egypt.

The social and religious world of the New Testament was shaped by the influence of Greek thought, Greek philosophical systems, and Greek moral values that had affinities with Christianity. However, as Copleston wrote, many years ago:

> Christianity was not, of course, in any sense the outcome of ancient philosophy, nor can it be called a philosophic system, for it is the revealed religion and its historical antecedents are to be found in Judaism; but when Christians began to philosophise [sic], they found ready at hand a rich material, a store of dialectical instruments and metaphysical concepts

and terms, and those who believe that divine Providence is operative in history will hardly suppose that the provision of that material and its elaboration through the centuries was simply and solely an accident.[1]

Greek converts had to learn that Christianity could not "be called a philosophic system" and that true wisdom was to be found elsewhere than where they had previously sought. Paul's words were both instruction and admonition for them:

> Has not God turned the wisdom of this world into folly? Since in God's wisdom the world did not come to know him through "wisdom," it pleased God to save those who believe through the absurdity of the preaching of the gospel. . . . For God's folly is wiser than men . . . (1 Cor 1:20-21, 25).

However, Christians did begin to philosophize. They sought not only God's wisdom, but in Christ, the "wisdom of God" (1 Cor 1:24). They sought to engage with and even to transform the "material" divine providence had placed at their disposal. Out of the many Christians who have sought and loved wisdom in every age, there are three whose unique insights can guide us in our own efforts to seek, to love, and to possess what God alone can give (cf. Wis 8:21). These three are Clement of Alexandria, St. Augustine of Hippo, and St. Louis Marie de Montfort.

CLEMENT OF ALEXANDRIA (C. 150–215)

A search of the calendar of the Roman Catholic Church for a feast day honoring Clement of Alexandria is, at least, disappointing. It is said that Titus Flavius Clemens Alexandrinus has been "on" and "off" the calendar of saints more frequently than any other friend of God. The reason is simple. Clement, the first Christian humanist,[2] the "pioneer of Christian scholarship,"[3] stands under a cloud of suspicion because of what are considered numerous errors or exaggerations attributed to him. However, he was looked upon as a saint by many ancient authors. Pope Clement VIII and Pope Benedict XII both removed Clement's name from

[1] Frederick Copleston, *A History of Philosophy: Greece and Rome,* vol. 1, pt. 2 (Garden City, N.Y.: Image Books, 1962) 250.

[2] Clement of Alexandria, *Christ the Educator,* Simon P. Wood, trans. (New York: Fathers of the Church, Inc., 1954) x, passim; also see Francis Hermans, *Histoire Doctrinale de l'Humanisme Chrétien* (Tournai-Paris, 1948) 4 vols.

[3] Johannes Quasten, *Patrology* (Utrecht-Antwerp: Spectrum Publishers/Westminster, Md.: The Newman Press, 1964) II:6.

the Roman martyrology, the latter because no evidence of a cult of veneration to him in any local church existed prior to the eleventh century.[4]

Clement was one who truly loved wisdom. Most probably born in Athens of pagan parents, his education included the study of philosophy, poetry, archaeology, mythology, and literature. He was "a Greek of the Greeks, steeped from infancy in the glory that was Greece."[5] As a seeker of wisdom who became a Christian, he set out to find the best teachers, traveling through Italy, Syria, and Palestine. Around the year 180, Clement settled in Alexandria to become the pupil of the priest Pantaenus, the first director of the famous Didascalia in that city. Clement has recorded his admiration for this universally acclaimed instructor:

> When I came upon the last [teacher]—he was the first in power—having tracked him out concealed in Egypt, I found rest. He, the true, the Sicilian bee gathering the spoil of the flowers of the prophetic and apostolic meadow, engendered in the souls of his hearers a deathless element of knowledge.[6]

Clement, lover of wisdom, found himself at home in Alexandria, first because of the rich instruction and, even, the gift of friendship he received from Pantaenus. He was also at home in a city that was the center of widely recognized intellectual and literary achievement. There were two outstanding libraries in the city, the Serapeion and the Museum. Philo, the famous Neo-Platonic Alexandrian Jew, had left a strong legacy of Hellenistic culture through his efforts to combine the teaching of the Old Testament with Greek thought. In such an atmosphere the Christian community turned to a deeper study of Christian teaching than prevailed in other regions. The catechetical school was marked by a spirit of liberal culture and humanism. It was conducted privately, "neither begun nor directed by ecclesiastical authority. . . . The pupils . . . were almost entirely Alexandrian Greeks, whether pagan or Christian, drawn from the more well to do members of society."[7] Both women and men—all "vigorous in [the] search for truth"—attended classes.

Following the death of Pantaenus, Clement became the leader of the school. The program was varied, including a rich study of the Scriptures, following the allegorical method of interpretation used by Philo. Other

---

[4] F. Cayré, *Patrologie et Histoire de la Théologie* (Paris-Tournai-Rome: Desclée & Cie, 1953) I:225.

[5] Clement of Alexandria, *The Educator,* vi.

[6] Cf. Quasten, *Patrology,* II, 5.

[7] Ibid., ix.

studies reflected Clement's conviction that there was an "integral relationship between all that was worth while in pagan literature and the new
Christian faith."[8] This attitude was shared by Origen (d. c. 254), Clement's
most famous pupil and successor in the leadership of the Didascalia,
who claimed that nothing of interest to the human mind was to be considered foreign to his classroom. Later, St. Catherine of Alexandria (d.
305), a brilliant scholar and now the patroness of philosophers, shared
this legacy.

Why, then, has Clement of Alexandria existed under a cloud of suspicion over the centuries? Three areas of his teaching can be identified
as suggesting an answer to this question. They are his teaching on
Christology, his system of Christian Gnosticism, and his understanding
of Christian perfection. Each of these points can be considered briefly.

Christology

The difficulty of accepting Clement's Christology as orthodox may
not be readily perceptible. He presented a doctrine of the Logos, "the
highest principle for the religious explanation of the world . . . the creator of the universe . . . the one who manifested God in the Law of the
Old Testament, in the philosophy of the Greeks and finally in the fullness of time in His incarnation."[9] The work that most fully and clearly
presents Clement's Christology is the *Paidagogos*. For Clement, Christ is
the preeminent teacher, the educator. This all-loving Word of the
Father, "anxious to perfect us in a way that leads progressively to salvation, makes effective use of an order well adapted to our development;
at first, He persuades, then He educates, and after all this He teaches."[10]

The words of one of the earliest known Greek Christian hymns
could have been written by Clement:

> Bridle-bit of colts untamed,
> Thou Wing of birds not straying,
> Firm rudder of our ships at sea,
> Thou shepherd of God's regal sheep. . . .
> O Jesus, our Christ!
> Milk of the bride
> Given of heaven,
> Pressed from sweet breasts—
> Gifts of Thy wisdom—

[8] Ibid., x.
[9] Ibid., 21–22.
[10] Clement of Alexandria, *The Educator*, 5.

These Thy little ones
Draw for their nourishment;
With infancy's lips
Filling their souls
With spiritual savor
From breasts of the Word.[11]

We ought not be surprised that Clement's intense poetical energy and enthusiasm in expressing his love for Christ can prove to be contagious!

Christian Gnosticism

Clement's Hellenistic education led him to develop the Christian faith into a system of thought with what was, for the day, a scientific foundation. Like Irenaeus of Lyons (d. c. 202) before him, Clement recognized the dangers in the false and heretical g*nosis* so prevalent in his age. Unlike Irenaeus, Clement did not totally reject the teachings of the various Gnostic systems. He recognized their errors in invoking a Christian vocabulary to teach their "redeemer myths," their esoteric understanding of human nature, and their claim that faith and knowledge were contradictory to each other. Rather than dismiss Gnostic teachings completely, Clement set up a "true and Christian *gnosis.*" He sought to place at the service of faith whatever treasures of truth were to be found in the various schools of philosophy that he knew so well. Clement's "Christian gnosticism" did carry signs of the esoteric teachings of the heretical Gnostics. He was not entirely free of attributing an exaggerated role to the function of knowledge in salvation, although he did seek to prove that faith *(pistis)* and knowledge *(gnosis)* rightly understood existed in harmony. Through this harmony, the perfect Christian, the true Gnostic is formed.

Clement's doctrine of Christian Gnosticism became a powerful tool against the heretical teachings of the Gnostics. Their subtle seductions were resisted by well-educated intellectuals whose reading of the Scriptures was informed and enlightened. Alexandria came to be recognized as the center of a culture that was "imbued with the Christian spirit." Christian Gnosticism, as Clement taught, avoided both the exaggerated "spiritualization" of the body, on the one hand, and the rejection of its "earthiness," on the other. The followers of these ideas fell into lives of luxury, debauchery, and vice. In and through Christ, the wisdom of God, the Christian Gnostic was called to envision the heights

---

[11] Ibid., 276–77.

of perfection. Clement's system, while marked by some suspicious traits, answered a need and provided a path that enabled many to become, to be, and to remain Christian.

Christian Perfection

Clement's idea of perfection for the follower of Christ was marked by the influence of Gnosticism and its notion of ascent from an earthly, "bodily" existence through increasingly spiritual aeons in their approach to an unknowable deity. Clement's doctrine presented the "moral" life as an *ascesis,* an ascension by which the human being passed through successive states toward perfection. He perceived each state as a "dwelling" in the "castle of the soul," every one characterized by a higher motive than the previous one: fear, first of all; then, faith and hope; finally, charity.

Clement made no distinction between women and men on the road to perfection:

> Let us recognize, too, that both men and women practice the same sort of virtue. Surely, if there is but one God for both, then there is but one Educator for both. . . . The very name "mankind" is a name common to both men and women. . . . Notice, too, that "sheep" is the general name used for the male and the female.[12]

He did, however, ordinarily divide Christians into two classes. Simple Christians possessed only a simple faith, common to all. They were not able to grasp the deeper meanings of Scripture, nor the more "mystical" doctrines of revealed truth. The perfect Christian, the true Gnostic, had a fully developed faith. Clement's insistence on the characteristics of the perfect—*apatheia,* charity, *gnosis*—seems to imply that there are two classes in the Christian life, one superior and the other inferior. With *gnosis* as the primary characteristic of perfection, contemplation not love, emerges as the final step in Clement's higher "religious knowledge" and union with God to which Christians are called.[13]

ST. AUGUSTINE OF HIPPO (354–430)

The second lover of wisdom in this "triumvirate" is much better known than Clement of Alexandria. Aurelius Augustinus was not only a

---

[12] Ibid., 10.
[13] Cf. Cayré, *Patrologie et Histoire de la Théologie,* 231–34.

lover of wisdom; he stands recognized as a teacher of wisdom, a philosopher in his own right. For Augustine, philosophy is wisdom that is love of supreme truth as the rule of life. His nine philosophical treatises, written over a period of ten years (386–395), set forth the fundamental principles of his theological reflections as well as of his preaching. Augustine's philosophy is theocentric, finding in God an explanation of all things.[14]

The Quest Begins

Augustine discovered wisdom in 373, at the age of nineteen, on reading Cicero's *Hortensius*. His account of this moment vibrates with all the enthusiasm of the experience:

> [W]hat brought me relish in this exhortation [of Cicero] was that I was excited and aroused and inflamed to love, seek after, attain, and strongly embrace, not this or that philosophic school, but wisdom itself, whatever it is.[15]

The *Hortensius* is extant only in fragments, but we know that it sought to encourage its readers to pursue wisdom through renunciation of ambition and pleasure, striving to "tame one's body as one would a wild horse."[16]

However, Augustine's taming of his wild horse had to wait. His discovery of wisdom coincided with his adherence to the Manichaeans. Persuaded by their eloquent and alluring claim to possess all knowledge, Augustine turned from a fruitless reading of the Scriptures to follow the Manichees. He was attracted, especially, by their rationalism, their rejection of the need for faith, their profession of a pure and spiritual Christianity that had no need of the Old Testament, and their radical solution to the problem of evil.[17]

Augustine's love of wisdom and persevering search to attain it runs like a leitmotif through the *Confessions*. He recalls that, at the time of reading Cicero's work, he realized that "the love of wisdom bears the Greek name, philosophy."[18] As his quest continued, he encountered Faustus, the Manichaean bishop, and began to ponder the difference

---

[14] Ibid., 711–13.

[15] Saint Augustine, *Confessions,* Vernon J. Bourke, trans. (New York: Fathers of the Church, Inc., 1953) 55–56.

[16] Garry Wills, *Saint Augustine* (New York: Viking Penguin, 1999) 27.

[17] Ibid., 110.

[18] Saint Augustine, *Confessions,* 55.

between true wisdom and the appearance of it. The soul is not necessarily wise because the face is handsome or the speech graceful!

> Rather, it is much the same with wisdom and folly as with edible and inedible food; both wisdom and folly can be presented in ornate or in plain language, just as both kinds of food can be served upon elegant or simple platters.[19]

Faustus, the man "who had been for so many people a deadly snare" became the one through whom Augustine was freed from his shackles. But the search for true wisdom and the persevering quest for God were not to be soon ended:

> I experienced the greatest wonder when I anxiously reflected on what a long time it was from my nineteenth year, at which time I had begun to be passionately interested in the pursuit of wisdom. . . . And here I was, now, in my thirtieth year, stuck in the same mire by my craving to enjoy the things of the present which eluded me and wasted my energy . . . .[20]

However, with every step forward, new light was given for this seeker to penetrate the mysteries of life and death.

> [W]ho else recalls us from the death of every error, except the Life which knows no death, the Wisdom which enlightens the minds that need it, yet itself needs no light, by which the world is governed down to the fluttering leaves on the trees?[21]

From Grace to Grace

Augustine's understanding of philosophy and of wisdom changed over the years, as he matured spiritually. In the nine philosophical treatises, his initial desire for God was expressed in ardent, almost feverish terms. His mind labored, confident that it could find the answers to the questions that tormented him. He was sure he could attain absolute peace and blessedness in this life, through his own striving. It was only later that he would grasp the role of what he came to call the "moral life" of piety and grace.

Augustine's earliest philosophical treatises resulted, in part, from the discussions he exchanged with those who were his young disciples and companions. With them, he realized that happiness was to be found

---

[19] Ibid., 110.
[20] Ibid., 151.
[21] Ibid., 170.

in knowledge of the truth, not in the possession of material goods, but in their balanced use. The wise are guided by providence beyond evil through an education that unites reason, authority, and the liberal arts to arrive at a perfect knowledge of God. Later treatises explore proof for the existence of God through the reality of one's own being, founded on eternal verities that come from the transcendent, immutable truth that is God.[22]

In each book of the *Confessions,* Augustine's journey to God is described through theological and philosophical reflections on an experience in his life or a problem with which he was grappling. Particularly noteworthy are the contents of book six, in which he recounts the struggles to come to terms with the demands of the wisdom he and his companions sought in their desire to live the "good" life. Finally, Augustine was to realize that only through the wisdom of God, Christ Jesus, "equal with God, God along with God, and at the same time one God," are we transformed from slaves into children of God.[23] Augustine's search for wisdom is the story of a soul that could be the story of every soul who is a lover of wisdom, a God-seeker. As Augustine's text reaches its climax, he celebrates the ways in which he has been led to God by God. The cry that springs from his heart is one that resonates in every human heart. "Late have I loved Thee, O Beauty so ancient and so new, late have I loved Thee!"[24]

In *The City of God,* Augustine invokes the wisdom that not only comes from Christ, but which is Christ, in order to read the story of the history of creation and with it the history of humankind.

> Seen in the light of Christian wisdom, the evolution of world history is a no less striking "confession" of the love and power of God than the sight of His creation, and the awareness of the wonders wrought by grace in the soul of His servant, Augustine.[25]

Divine Wisdom is the bond that unites the society whose ideal Augustine traced for those who would come after him.

> No wisdom is true wisdom unless all that it decides with prudence, does with fortitude, disciplines with temperance, and distributes with justice

---

[22] Cf. Cayré, *Patrologie et Histoire de la Théologie,* 711–13, as a source for much of this material.

[23] Cf. Saint Augustine, *Confessions,* book ten, passim.

[24] Ibid., 297.

[25] Saint Augustine, *The City of God, Books I–VII,* Demetrius B. Zema and Gerald G. Walsh, trans. (New York: Fathers of the Church, Inc., 1950) lxxxii.

is directed to that goal in which God is to be all in all in secure everlast-ingness and flawless peace.[26]

The Fullness of Wisdom

Augustine was well acquainted with the history of philosophy. During the early years of his teaching career, he had read and known by heart many works of the philosophers. His favorite authors were Plotinus and Porphyry; his favorite school, Platonism. The "fundamental lines" of Augustine's philosophy have been identified as interiority, participa-tion, and immutability.[27] These characteristics, along with his basic doctrinal principles, led to the designation of "three wisdoms": natural, theological, and mystical.

Natural wisdom can be defined as "a simple, rational support for Christian wisdom." It leads to a partial knowledge of God as the basis for religion and the moral life through the world, and all that is true and good. All things are to be considered in relation to their first, supreme Principle.[28]

Theological wisdom is the foundation of the Christian life. It is love for revealed truth, leading to God and bringing the supernatural life to fruition in the baptized. Theological wisdom is nurtured by the virtues of faith, hope, and charity and becomes a kind of synthesis of them in the soul, bringing forth the gifts of the Holy Spirit and a life according to the eight beatitudes.

Mystical wisdom is the highest gift of the Holy Spirit. Through this gift, the soul comes to "repose" in contemplation and even now tastes eternal life. Augustine identified three biblical persons who exemplified this wisdom: Rachel, the sister of Leah; Mary, the sister of Martha; St. John, Apostle and beloved disciple.[29] In these three, Augustine found lovers of wisdom who had sought God through contemplation and had found him in the perfection of a love that united them to God. At the end of the quest for wisdom, faith gives way to vision and God is all in all.[30]

---

[26] Saint Augustine, *The City of God, Books XVII–XXII*, Gerald G. Walsh and David J. Honan, trans. (New York: Fathers of the Church, Inc., 1954) 232.

[27] Quasten, *Patrology*, 4:407.

[28] Cf. Cayré, *Patrologie et Histoire de la Théologie*, 739–43.

[29] Cf. ibid., 742.

[30] Saint Augustine, *The City of God, Books XVII–XXII*, 505–06.

ST. LOUIS MARIE GRIGNION DE MONTFORT (1673–1716)

Our third lover of wisdom is not a Father of the Church, although he represents a time and a tradition of spirituality marked by a return to the patristic era. After St. Augustine's death, a gradual disinterest in, misuse, and finally abandonment of the patristic legacy prevailed in the development of Christian thought. With the Renaissance, a rediscovery of the classical achievements of Greece and Rome occurred. The neo-classicists continued this movement with their retrieval of the early Fathers, particularly Augustine. When Erasmus (d. 1536) introduced his friend, Thomas More (d. 1535), to the Fathers of the Church, he planted the seeds that were to blossom later in the Oxford Movement.

During the Protestant Reformation Augustine emerged as the authority to be relied upon for an unmasking of the real and alleged failures of Rome. Although frequently read selectively and interpreted out of context, Augustine provided what the Reformers needed as they sought to restore Christianity to the norm established by Scripture alone.

The Wisdom of Reform and Renewal

St. Louis Marie de Montfort was the "last master of the Bérulle School."[31] Begun in response to the call of the Council of Trent (1545–1563) for a desperately needed renewal of the Catholic Church, this school launched a major pastoral reform in seventeenth-century France. The women and men who were spiritual leaders[32] of this school found inspiration and affirmation of their efforts in a return to the wellsprings of Christian life and thought: the Scriptures and the Fathers of the Church.

Montfort was not a "lover of wisdom" in the line of either Clement of Alexandria or Augustine of Hippo. In the first place he has often been regarded as a "maverick" because of aspects of his personality that seem eccentric. An early biographer notes that "worldly men deemed him strange, good men unique," more to be admired than imitated.[33]

---

[31] Cf. Raymond Deville, *The French School of Spirituality,* Agnes Cunningham, trans. (Pittsburgh: Duquesne University Press, 1994) 189.

[32] The spiritual leaders of this school are well known: Madame Acarie (whose *cercle* counted more than one hundred members, including St. Francis de Sales and St. Vincent de Paul), Bérulle, de Condren, Olier, Madeleine de Saint-Joseph, St. John Eudes, Marie de l'Incarnation Guyart, St. John Baptist de la Salle, and St. Louis Marie de Montfort.

[33] *Jesus Living in Mary, Handbook of the Spirituality of Saint Louis Marie de Montfort* (Bayshore, N.Y.: Montfort Publications, 1994) 754.

He has been described as an "extraordinary saint," a "spiritual man," a "man of the streets." Because Montfort found it impossible to conform to an established framework, his spiritual director at the seminary perceived his actions as excessive and extravagant behavior. The "poor and little ones" would always see him as a man of God. For others, he would be, if not a deliberate troublemaker, an enigma. Few realized that his sensational actions and apparent anti-conformism flowed out of his overwhelming sense of the "absoluteness of God and the radicalness of the Gospel."[34]

Montfort's quest was the result of what has been identified as the charism of apostolic wisdom, a gift of the Holy Spirit. It bore fruit in his life as "the Wisdom of love" and found expression in a "surprisingly dynamic apostolic quality." Throughout his life he could be described in the words of one of his friends from seminary days:

> I have to say that he reflected the strength and impetuosity of the new wine of the Holy Spirit from which apostles come. He is a madman, and a fool in the eyes of men, but a sage in the eyes of God. . . . His complete innocence is linked to his limitless penance, continuous prayer, and limitless mortification. He offers the Holy Spirit a pure heart ready for each and every undertaking.[35]

Underlying Montfort's spirituality was the mystery of the Cross.

The Wisdom Cross of Poitiers

As a newly ordained priest Montfort was assigned as chaplain to a large poorhouse in Poitiers, France. The primary means he chose to bring about reform in a deplorable situation was a prayer group of disabled women residents. The group was named Wisdom, and in its meeting room Montfort placed a large cross that he had designed. An inscription on the cross proclaimed the glory of suffering with Christ, Eternal and Incarnate Wisdom crucified. This cross became known as the Wisdom Cross, the Cross of Poitiers. For Montfort, the Wisdom Cross was to be a reminder that we are redeemed through the victorious cross of Jesus Christ. Paul's words to the Corinthians (1 Cor 1:23-25) were Montfort's message to those to whom he ministered. The symbols on the cross represent Infinite Love Incarnate in the Sacred Heart of Jesus along with an understanding of Our Lady's role as a model for us of sharing in the sufferings of her Son. The inscriptions

---

[34] Ibid., 761–63.
[35] Ibid., 138–39.

on the cross are a summons to the Lord's disciples to share these sufferings, like Mary and with her, through humility, submission, patience,
and obedience.[36]

Montfort's entire spiritual journey can be described as a passionate
search for Wisdom. It was pursued in the light of the mystery of the
Cross and its proclamation that the Wisdom of God is contrary to the
wisdom of the world. Wisdom, for Montfort, was Jesus Christ, Eternal
Wisdom incarnate and crucified. He knew the price to be paid to possess this treasure.

> Keep on praying, even increase your prayers for me; ask for extreme
> poverty, the weightiest cross, abjection and humiliations. I accept them
> all if only you will beg God to remain with me and not leave me for a mo-
> ment because I am so weak. What wealth, what glory, what happiness
> would be mine if from all this I obtained divine Wisdom, which I long for
> day and night![37]

Through love for Jesus Christ, Montfort developed an understanding of
Wisdom that has been the source of theological and spiritual enrichment in Christianity.

To Jesus through Mary

The cross was not the only object of Montfort's contemplation of
Wisdom. Louis Marie also perceived Wisdom through a contemplation
of beauty. This was especially so when he considered Mary. In the incarnation, Montfort insists, God experienced beauty in a new way by
creating Mary, filling her with grace, and placing in her "unutterable
marvels and beauties." To contemplate the beauty of Mary is to discover
the beauty of her son, Jesus. To contemplate Jesus is to know and love
Wisdom Incarnate.

Montfort's best known work on Mary is entitled *True Devotion to
the Blessed Virgin*. It was addressed to the "poor and simple," although
theologians have found it "profoundly rich and contemporary."[38] The
heart of true devotion to Mary is the conviction that, because Christ
came into the world through Mary, it is through her that he must become known and reign more and more. Montfort understood that true
devotion to Mary is "a slavery of love." This slavery means living ever

---

[36] A detailed description of the Wisdom Cross can be found in ibid., 260–66.

[37] Letter to Marie Louise Trichet, cofoundress with Montfort of the congregation of
the Daughters of Wisdom; cited in ibid., 767.

[38] Ibid., 1214; cf. 1209–29.

more deeply a life of faith and loving Mary as mother and mistress, so as to be filled with the Spirit for "God Alone."[39]

Montfort's devotion to Mary matured, especially, through his preaching of the faith in retreats and missions in parishes, barracks, houses of prostitution, and town squares. He called for total consecration of oneself to Jesus through Mary and the interior practice of doing everything through Mary, with Mary, in Mary, and for Mary, in order to do it more perfectly through Jesus, with Jesus, in Jesus, and for Jesus. The first effect of his preaching was in his own life: "I have her image carved within me. I carry her in the center of my being."[40]

From the Incarnation to the Trinity

The foundation of Montfort's Marian doctrine is her role in the Incarnation. Through her consent to God's proposal, Mary became Our Lady of Wisdom. Through God's grace, she was united in an "ineffable" way with each person of the Trinity. Mary and the Father have the same Son. Mary and the Son share total dependence on the Father. The Holy Spirit communicates to her the infinite loving which binds Father and Son together.

Indeed, for Montfort, the Trinity is a mystery of love.[41] The Father is "originating Love"; the Son is born out of the fullness of the Father's love, for the Word is the "breast of the Father," the "Art of the Father," the "unsurpassingly beautiful" expression of the Father. God the Holy Spirit, infinite Love "chose to make use of Our Lady . . . in order to become actively fruitful in producing Jesus Christ." And Montfort exclaims to the Spirit: "All the saints who have ever existed or will exist until the end of time, will be the outcome of your love working through Mary."[42]

CONCLUDING REFLECTION

In every age the disciples of Jesus Christ are called to be lovers of wisdom, seekers of God. In Clement, Augustine, and Montfort we find three unique exemplars of the multiple ways in which grace effects such love and seeking in our lives. Clement urges us to seek truth wherever it is to be found, in the achievements and challenges of human culture

[39] "God Alone" is often cited as Montfort's personal motto; cf. ibid., 697.

[40] Ibid.

[41] For a full presentation of Montfort's Trinitarian doctrine, cf. ibid., 1177–90.

[42] Ibid., 1186.

and learning. Augustine encourages us to look at creation, human friendship, the history of humankind and of the world through the eyes of faith, hope, and charity, the three theological virtues. Montfort would caution us not to abandon the cross and not to forget Mary. Because her role was willed by the Trinity, neglect of her can lead to a lessened living of the mystery of the Trinity in the Christian life. That can mean only that there would be fewer Christians to love wisdom and to seek God through Jesus Christ, Eternal Wisdom.

To each of us is given the grace to love wisdom and to seek God as women and men of our own time and place. Our concerns and interests are unique in ways with which we are familiar, even when we are uncomfortable with them. Can we do less than those who have gone before us in facing the challenges that would enable us to grow as lovers of wisdom, as seekers of God? Clement, Augustine, and Montfort challenge us to think carefully before we answer this question!

8

"Charged with the Grandeur of God"

The Created World as a Path to Prayer

ANTHONY J. GITTINS, C.S.SP.

IMAGE OF GOD

On the wall of my study hangs a carving of the crucified Christ in faux-ebony, purchased in Tanzania a dozen years ago. It is some thirty inches high and its African features are beautifully crafted. There is no crucifix behind it and no screws are visible, so the Christ figure appears to hang suspended in midair, the neutral, featureless wall providing merely a background. This three-dimensional image is part of the environment. It is always there, even when no one else is, even when it is unnoticed.

The African Christ continues to touch my life. Sometimes when I raise my eyes, I notice again the serene-faced one, eyes downcast, and arms open wide in welcome embrace. Sometimes I can imagine the cross against which his body writhes and the inaudible groans. More often than not, when my attention is caught by this beautiful artifact, I am simply aware of the sheer *Africanness* of this body: attenuated limbs, etched rib-cage, high cheek bones, and dark, dark skin. If the Muse is present I might reflect that since all humanity is made in God's image, and since the Incarnate One is God brought down to earth in human flesh and blood, then unless I can look on other faces like this—African or not—and see images of God every bit as striking, then I remain impoverished because my idea of God is constrained by images derived from my own culture or limited by my own previous encounters.

GOD'S REVELATION

Addressing the Catholic Theological Society of America,[1] Michael Amaladoss, s.j., developed a rather similar reflection. He was speaking of Asia, but his words have global relevance:

> Many Asian theologians accept that God's salvific encounter with the members of other religions takes place in and through these religions, not in spite of them. The other religions therefore have a role in the salvific plan of God for humanity . . . (1).
>
> The Asian Bishops . . . in 1974 . . . assert[ed] that dialogue with the believers of other religions "will reveal to us what the Holy Spirit has taught others to express in a marvelous variety of ways" (2).
>
> [In Assisi in 1986, Pope John Paul II said that] "every authentic prayer is from the Holy Spirit" [and in *Redemptoris Missio,* 1990, para. 28, that] "[t]he Spirit's presence and activity affect not only individuals but also society and history, peoples, cultures and religions." [That] invitation took for granted that the believers of other religions can be in contact with God in prayer and that their prayer is effective. . . . The other religions therefore are not merely human efforts, but embody an ongoing divine-human encounter. . . . Every divine-human encounter is salvific, [and] all religions mediate or facilitate salvific divine-human encounter (2).

Amaladoss develops his theme in the following points: that most people die without encountering or being fulfilled by the Church (3); that we have no business to claim that other people's religious experiences are illegitimate (4); and that we should acknowledge that Jesus the Christ does seem able to inspire people, quite apart from the mediation of the church (6).

Such recent thinking on the salvific value of religions other than Christianity is thought provoking, but still leaves an important issue unexplored. It is not enough to assert that God is self-revealing or that God can be known in some manner through any of the world's religions,

---

[1] Michael Amaladoss, "Pluralism of Religions and the Proclamation of Jesus Christ in the Context of Asia," *Proceedings of The Catholic Theological Society of America* 56 (2001) 1–14. Numbers in parentheses are page references.

unless we also try to identify more particularly just how this might come about. The Church has spoken of Divine revelation happening in a way known only to God, or in a mysterious way.[2] However, when we come to concrete cases, this is simply not adequate. So let us pursue the matter.

Short of direct and personal revelation, or the importation of messages from outside, people can only be in contact with God through the thoughts, images, objects, and relationships found in their own actual worlds. Therefore, we would do well to explore some aspects of those worlds, for God is not revealed exclusively to people who share a single social heritage (and certainly not something simplistically and arrogantly labeled "Western Culture"). God—Creator, Christ, and Comforter—is for all times, all people, and all places: for yesterday, today, and tomorrow; for Jew and Greek, slave and free (Gal 3:28); for the whole world, the universe, the cosmos.[3]

The Jewish-Christian tradition asserts that humanity—every person, male and female—is made in the image of God. If this is so, then the more authentic encounters each of us has with other human beings, the more we will be exposed to the many faces—the multi-facetedness—of God. Similarly, the more insular our experience and limited our encounters, the less we know *of God* and the less we *know God.* It is deeply ironic that some of those who have the most superficial encounters with others, or the least experience of I-Thou relationships, are among the very people who presume to speak about, and even on behalf of, God. "What should they know of England, who only England know?" asked the British expatriate Rudyard Kipling. What could we know of God if we only know a handful of familiar images?

This essay is quite simple. Having started with a reflective glance at an African carving, it proceeds by looking at some other faces of God (not carved or painted, but fashioned from words) drawn from across the length and breadth of Africa. It will continue with a reflection on our common need to expand our cultural horizons and be edified by hitherto unimagined aspects of God, and it concludes with some observations about cross-cultural enrichment and mutually enlightening conversations. But first, a belated foreword.

---

[2] *Gaudium et Spes,* 10, 15, 22. *Ad Gentes,* 9.
[3] *Halleluia for the Day: An African Prayer Book,* ed. Anthony J. Gittins (Liguori, Mo.: Liguori/Triumph, 2002) xi–xvii.

TREASURE ENOUGH FOR ALL

Within a few short years, the Church's center of gravity will have so palpably shifted southwards to the equatorial regions of Africa that everyone will know it, though some will be surprised and perhaps offended. African faces and African images of God (not to mention many other faces and images) will become increasingly part of Christianity's global repertoire. It would be more than tragic if conservative and unimaginative Christians failed to be enriched by such accessible treasures. After all, Europe exported its own cultural images of God and Christ, expecting others to be enriched and grateful; if cultural borrowing and lending can proceed in one direction, why not in the opposite direction?

There are several reasons for a lack of religious reciprocity. Some people do not realize that their own ideas and images of God are cultural constructs. Some seem to think that their *idea* of God is itself a true image or representation of God.[4] Some again seem to think that since they have inherited a religious tradition that their own forebears transmitted to Africa, it therefore follows that Africa is a recipient of, rather than a creative contributor to, the Christian tradition. Yet even in the so-called first world, people have come to acknowledge the serious limitations of images of God and Christ that grew from the soil of Western cultures (God as mighty warrior, plenipotentiary, or implacable judge; Christ as king, shepherd, vine, lamb). Though each of these may contribute analogically to our understanding of the Transcendent, they are all limited and some have created serious problems for would-be translators. This may be equally true of images drawn from other cultures, yet some of those are no less striking than the best of our own. The very unfamiliarity of images from other cultures might actually inspire and remind us once again that God is simply not reducible to our images of God, however suggestive they may be. We cannot understand God; but if we endeavor to *stand under* what we cannot *understand*, we may be given a new perspective.

---

[4] "Not a single missionary [in Latin America] was aware of the fact that the God the church proclaimed was a cultural image, developed in a syncretism of biblical, Greco-Latin and barbarian material, and not actually God, who always transcends language and representation. In the same fashion, the divinities of the Indian cultures were only representations of the mystery of God, of course, and not actually God. The essence of idolatry is the identification of the reality of God with the image of God produced by a culture. And the missionaries blithely identified their image of God with God—just as the natives did." Leonardo Boff, *The New Evangelization* (Maryknoll, N.Y.: Orbis Books, 1991) 117.

"The Journey of the Magi," T. S. Eliot's powerful, bleak poem, challenges all who have become comfortable with their understanding of God. The Magi returned from Bethlehem "no longer at ease in the old dispensation." The people they encountered on their return home were now perceived as an "alien people clutching their gods," while the Magi had been deeply disturbed and transformed by their experience. How can God possibly succeed in transforming us unless we open own hands and hearts, share our gifts, and receive treasure we could never have imagined? Some African images of God may help us to reflect on our own familiar gods. Then, rather than clutching them even tighter, we might become more open to other gifts, more able to recognize Emmanuel: God-with-us.

In recent years, Christian theology has begun to acknowledge not simply the existence of other religious traditions and their fundamental right to respect, but their own intrinsic riches. Christians committed to interreligious dialogue have urged their fellows to understand that such traditions contain "seeds of the Word," that they are salvific for all those who follow the dictates of consciences formed by their own religious traditions. Even if we limit ourselves to Africa in what follows, we can surely be graced and enlightened by contemplating some of these riches.

AFRICAN IMAGES OF GOD

In East Africa the Maasai range across parts of Kenya and Tanzania known to the outside world by such names as the Rift Valley and the Serengeti Plain, a vast expanse of grassland teeming with wild life. Here is one of their prayers:[5]

> Creator God, we announce your goodness because it is clearly visible in the heavens where there is the light of the sun, the heat of the sun, the light of night. There are rain clouds. The land itself shows your goodness, because it can be seen in the trees and their shade. It is clearly seen in water and grass, in the milking cows, and in the cows that give us meat. Your love is visible all the time: morning and daytime, evening and night. Your love is great. We say, "Thank you, our God" (1).

Reading this in the American midwest, the northern expanses of Canada or a European city may be unsettling; it may also provide a

---

[5] All the African prayers can be found in *Halleluiah for the Day*. The numbers in parentheses are page references.

helpful reminder of the sheer variety of the earth's bioregions and a telling glimpse of the beneficent presence of the Creator. For example, in West Africa among the Kissi and Limba people of Northern Sierra Leone, nature has carved spectacular sculptures: smooth-faced rocks of massive proportions rise behind the villages like towering scenery. Over in Nigeria's far north, the villages of the Mumuye people are scattered across bleak and boulder-strewn terrain, more moonscape than landscape. In Tanzania's Olduvai Gorge, a treasure-trove of hominid fossils, erosion has exposed geological strata and left the earth scoured and scarred. To walk in worlds like these is to feel self-consciously tiny and inspired by their awesomeness: a prelude perhaps, to a moment of religious insight; but to have been born and raised amid such physical geography is to be invited to incorporate its significance into one's worship. A prayer of the Rozwi people of South Africa combines such inspiration with a degree of familiarity:

> O Great Spirit!
> Piler-up of the rocks into towering mountains!
> When you stamp on the stones
> The dust rises and fills the land [. . .]
> You are the one who calls the branching trees into life:
> > you make new seeds grow out of the ground
> > so that they stand straight and strong.
> You have filled the land with people (4).

The world of traditional pastoralists and hunters in Tanzania offers a dramatic contrast: the Chagga know the difficulty of eking out a living and their need of any help they can get. Not surprisingly, their image of God is of the creator and sustainer of the actual world in which they live. As they prepare a bull, ritually offered as an apotropaic sacrifice (though the sacrifice serves to gather the community in festive sharing of beef, as much as to ward off evil), their earnest prayer goes like this:

> We know you God, Chief, Preserver,
> You who united the bush and the plain,
> You Lord, Chief, the Elephant indeed.
> You have sent us this bull which is of your own fashioning.
> Chief, receive the bull of your name.
> Heal [this person] to whom you gave it, and his children.
> Sow the seed of offspring within us, that we may beget like bees.
> May our clan hold together, that it be not cleft in the land.
> May strangers not come to possess our groves.
> Now Chief, Preserver, bless all that is ours (52).

For those raised or living in the concreteness of the modern city, this image of the unseen God may be as inspiring as any of our more familiar ones. It may perhaps stimulate us to seek out more word-pictures used by people living in environments very different from our own, and even to extend the limited range of our own God-images. Here, for instance, is a strong and earthy metaphor, embedded in a much longer prayer of the Pygmy people of Zaire:

> O God, thanks!
> You, my termite heap on which I can lean,
> From which come the termites that I eat; . . . (57).

This is the prayer of a corporate kinship group that gathers with its food-offerings, only too aware of the scarcity of food, of the fact that it is God's gift, and of their own responsibility to make return offerings to the God who gives. God is like the multivocal termite mound: massively visible against the surrounding terrain; complex and intricate in form and feature; fortress, home, and haven to uncountable living ones; producer of durable building material. The people acknowledge God's resource-rich presence in their midst, giving thanks to the generator of all their sustenance. For those of us born to abundance and used to plenty, this can be a wholesome reminder of our own dependence; for those whose food comes prepackaged from supermarkets, it may give pause to reflect on things we take for granted.

In almost a decade of living among the Mende people of Sierra Leone, West Africa, I began to learn the virtue of interdependence. But it was only years later that I discovered some of their indigenous prayers. The Mende speak with touching unselfconsciousness to a God of inclusion, a God of the community:

> O great God, help us, our families
> The men of this town, the people in general,
> our pregnant women,
> our daughters and our children especially.
> May our town progress.
> May our land progress.
> May our wives not die a mysterious death
> and our men not fall from palm trees.
> Help to see that all these things do not occur (60).

Another face of God is revealed in prayers from the Samburu of East Africa, who feel perfectly able to identify their immediate needs and to bring them before God with no sense that they are importuning the

Creator. Prayer after prayer concludes with God, far from being irritated or impatient, responding in a favorable, kindly, and delightful way:

> My God, guard us
> in the narrow and deep valleys full of dangers,
> and in the plains without end,
> and in the fords we cross small or large.
> And God said, "All right!" (62).

Or again:

> God, save us, God, hide us.
> When we go to sleep, God do not sleep.
> Tie us around your arm, God,
> like a bracelet.
> God, look on us with a countenance that is happy.
> Hit us with the black cloud of rain.
> God, give us what we ardently desire
> in regards to children and to cattle.
> God, do not make our land barren.
> God, give us places where there is life.
> God, divide us fairly into dead and alive.
> And God said, "All right!" (94).

To feel free enough to ask the Creator to assure the well being of a community composed of the unborn ("give us children"), the living and the dead ("divide us fairly"), and supported by rain for crops and cattle for food and exchange, is to be deeply blessed already. The God who so readily says "All right!" to humanity's inspired requests, is truly Emmanuel, God-with-us, and neither tyrannical nor capricious.

An introspective prayer from Ghana, quite realistic about mortality, nevertheless expresses quite beautifully the hope-filled trust of God's human creatures:

> Come, Lord, and cover me with the night.
> Spread your grace over us as you assured us you would do.
> Your promises are more than all the stars in the sky;
> your mercy is deeper than the night.
> Lord, it will be cold.
> The night comes with its breath of death.
> Night comes; the end comes; you come.
> Lord, we wait for you day and night (96).

These, in brief, are some of the myriad African faces or images of God, and the stuff of African prayer and reflection. We can surely learn from them and pray with them. We could also seek them out, not only in a thousand cultures in Africa, but in thousands more elsewhere.

GEOGRAPHY AS THEOLOGY

Metaphors, similes, images of God: these are most often crafted from the raw materials of nature or culture. Some unfamiliar images already disclosed may help to open our eyes and turn us to a more intentional contemplation of God's many faces. African prayers are but one kind of cross-cultural resource. There is another, and it is all around, though largely untapped by most of us except, perhaps, on very special occasions. It too may stimulate our prayerful contemplation.

There must have been times in our lives when the vision of a lustrous moonrise or shimmering sunset, the view of spectacular snow-capped peaks on a rugged horizon or the sight of a new-born and freshly-washed baby, have inspired us to thoughts of the Creator. There may have been other moments when the pure melodic line of a Beethoven sonata or a Bach partita has swept us up and far away from the mundane. Such tastes of transcendence or moments of unity with nature can be authentic religious experiences, though for people caught in the daily humdrum and stress they are only too rare. Yet we live in an age of travel, tourism, and transience, and we move with some ease across the face of the earth, and the world itself is a theology book, or at least an illustrated essay about the Creator. How sad if we are too preoccupied to know what we are missing. If we had the eyes to see, we might experience God in the ordinary as well as in the extraordinary.

The physical world itself is invested with levels of meaning: not just literal but metaphorical, allegorical, poetic meaning. In ancient times important events were linked to the geographical environment or to the rhythm of the seasons. Where people do not oppose and separate the social and the theological, it may be relatively easy to discern *sacred* geography and *sacred* time and space. If we come from a tradition more prone to opposing and separating, it may be much more difficult to perceive (or retrieve) the sacred in the secular. Yet it is surely worth the effort, for an active imagination (or eyes of faith) can interpret the world around us as an aspect of the Creator. This is no crude pantheism or panentheism but quite consistent with contemporary creation-theologies.

No individual can possibly see the whole of creation, any more than a single bioregion can possibly represent its variety. Travelers, theologians, and all who reach out to God can learn from other people and other worlds. Just as a Netsilik (Eskimo) or a Sami (Lapp) is attentive to winds and storms, though very differently from a Pacific Islander or resident of "Tornado Alley," so a person raised in the shadow of a towering peak or dwelling in a lush valley, living by the bountiful ocean or in a burning desert, cannot but attend to the environment, read its daily messages, and try to live in respectful symbiosis. For such people, theological ideas and metaphors are significantly shaped by actual worlds. Thus it is sometimes said that traditional religions are *embedded* in the culture rather than *institutionalized*, as Christianity has tended to become. Traditional religion is much less opposable to, or separable from, so-called secular life.

Many post-moderns have lost respect for the world in which they live. Not surprisingly this is accompanied by a degree of theological agnosticism, egocentrism, or airy indifference to ecology and environment. The metaphors of recent scientific knowledge do not encourage a developing relationship with the Creator, but offer impoverished images of blind watchmakers or of attractive but mathematically generated fractals. If we are ignorant of the phases of the moon or the formation of clouds, the course of the sun or the character of the seasons, such images can have no theological significance for us. If we are impervious to the presence of the Creator, they cannot evoke in us a sense of relationship. Even if the Great Barrier Reef, the Amazon rainforests, or Kilimanjaro's majesty are still awe-inspiring, who lives close enough to be awed, and who has not already been desensitized by the pabulum of the television travelogue?

By contrast, in many cultural traditions, creation is understood as the Creator's communication, and people know that their efforts to maintain vital relationships with Divinity are the measure of their own survival as communities. Monuments built upon significant sacred sites explicitly reinforce the bond between the life of local people and the life-force of the Creator or abiding spirit. How tragic that Western architecture, far from exhibiting respect for or deference toward the land, so often presumes first to gouge and reshape it, and is then vaingloriously superimposed upon it. Indeed, some Western-generated technologies are created explicitly to strip and to scar, to sterilize and rape the earth. Meanwhile there are still people who cherish the earth, not only as their home but as the body of God.[6]

---

[6] Sallie McFague, *The Body of God: An Ecological Theology* (Minneapolis: Fortress Press, 1993).

CULTURE, GEOGRAPHY, CONTEMPLATION

The world may be a theology book, largely unread by many literate people. Paradoxically, some people can read with skills literacy never dreamed of. Perhaps the reflections in these pages will sensitize us so that our daily encounters may become the stuff of contemplation. Cannot the people we meet, the places we visit, and the panoramas we discover be brought to prayer, stimulate our imagination, and lead us to meditation on the mystery of God? The Africa prayers considered here are only a fraction of the riches to be found in every culture, while the observations on the world as a theology book are simply instances that can easily be multiplied.

Like a faded sepia photograph, every representation of God is a pale reflection. Yet a treasured photograph can stir memories of another time, or place, or experience. Humanity needs memories, not to root us to the past but to evoke forgotten moods and motivations. Those who went before us are so much more than any single representation can capture, so a range of images is much more helpful than simply one, however subtle or luminous. Our painters and poets, storytellers and sculptors, ensure that the many faces of kings and queens, of presidents and statesmen, will continue to be revealed to us. Their legacy is our cultural heritage.

But we have a theological heritage, too. St. Paul assured the Corinthians (1 Cor 13:12) that we can only see, as it were, through a glass darkly (or a dim reflection in a mirror). Nevertheless, we can indeed glimpse the face of God if we look carefully and widely, for God's glory and God's shadow are all around. If Wilfred Cantwell Smith[7] and Leonardo Boff are right to warn us against reducing God to a single image that can so easily become an idol, they are also correct to challenge us: we must extend our perspective and expand our view. A conscious effort to do so might provide a less impoverished view of who God is, and generate in us a more appropriate response to the God who calls us to life and to love. As we look upon the Creator's work in nature and in culture, we might recall those soaring words of Gerard Manley Hopkins, from which the title of this piece comes:

[7] Wilfred Cantwell Smith, "Idolatry in Comparative Perspective," in John Hick and Paul Knitter, eds., *The Myth of Christian Uniqueness* (Maryknoll, N.Y.: Orbis Books, 1987) 53–68.

The world is charged with the grandeur of God.
It will flame out, like shining from shook foil;
It gathers to a greatness, like the ooze of oil . . . .[8]

---

[8] Gerard Manley Hopkins, "God's Grandeur," in *Immortal Diamond: The Spiritual Vision of Gerard Manley Hopkins* (Doubleday, N.Y.: Image Books, 1995) 38.

9

The "Myth of the Garden" and Spiritual Ministry in Postmodern America

MARY FROHLICH, R.S.C.J.

INTRODUCTION

In the video *Powers of Ten,* the camera begins by focusing on a young couple lolling on the grass in a park by Chicago's lakeshore. It next moves upward ten feet, then one hundred feet, then one thousand feet, and so on until our view is so many trillions of miles out into the universe that even the galaxy containing the earth is less than a speck. Finally, the progression reverses and—with lightning speed—we find ourselves returning from the far reaches of the universe to the young couple still relaxing by the lake. Although the point of the video is probably to relativize our natural close-to-the-nose perspective on the world, a somewhat different imaginative reflection on the video can also open up reflection on the challenges of spiritual ministry in the United States at the dawn of the twenty-first century.

The video does not give us any background on the young couple. Let us imagine, however, that they are twenty-something office workers who grew up in other states and have only lived in Chicago for a couple of years. They are not married, but have an apartment together in a nearby high-rise. They do not feel a strong relationship to the city or even to the park in which they are resting, other than that this is a green and open space that they can get to easily. At work, on the streets, and in social settings, they mix with people of every conceivable nationality, ethnicity, and religion. Although they do not read newspapers, through television and radio they feel inundated by the "wars and rumors of war" that are afoot in the world. As for church, they go occasionally with a vague feeling of nostalgia and a well-concealed hope that there might be something for them there. They go away not sure that there is; yet, a few weeks or months later, that feeling overtakes them again and they are back. What have we to offer them?

Obviously, there is much more we would need to know about such a couple: their ethnic heritage, family ties, personal aspirations, religious history, etc. Yet even without these, there are certain diagnoses we can make about their spiritual needs. They need a deep-rooted sense of identity and at-homeness that can bear the continual pressure of diversity and change; they need affirmation and guidance in fulfilling their longing for intimate human relationship; and they need a framework of meaning within which to make some sense of both the disturbing daily news and the *Powers of Ten*. And, very little of this is given to them in any simple way. Although they may not know the word "postmodern," they live the reality of having to continually create and recreate identity, community, and meaning in the midst of multiple inputs and flux on all levels. They live the reality of what Robert Schreiter named as the "de-territorialized, hyperdifferentiated, and hybridized" relation to context that is typical of the postmodern world.[1]

What have we to offer them? This essay is a modest attempt to explore how some of the images and myths enshrined in the biblical tradition may be able to be re-appropriated for ministry in such a world. As Philip Hefner has noted, the capacity and need for myths is built deep into our nervous systems, which innately strive to produce pictures and models of the world in order to gain some purchase on overwhelming amounts of incoming data. In particular, myths provide the psyche with a dramatic structure that places present actions in the context of a valued past and a desired future. Hefner suggests that, in essence, the Christian myth proposes that the "way things really are" is that the world is a place in which we really belong.[2] Yet this is exactly

---

[1] Robert J. Schreiter, *The New Catholicity* (Maryknoll, N.Y.: Orbis Books, 1997) 26–27.
[2] Philip J. Hefner, *The Human Factor: Evolution, Culture, and Religion* (Minneapolis: Augsburg, 1993) 12.

what is so deeply challenged by the "deterritorialized, hyperdifferentiated, and hybridized" world in which so many people live today. In view of this, I will focus specifically on how "the myth of the garden" as an archetypal story of at-homeness, wholeness, and hope has functioned in a variety of past contexts, and may yet have new riches to offer today.

THE BIBLICAL MOTIF OF THE GARDEN

Rather than attempt a complete study of the biblical use of garden imagery, here we will engage in a reflective exploration only of three of the most evocative instances. The three are: the Garden of Eden in Genesis 2–3, the Garden of the Woman in the Song of Songs, and the Garden of Passion and Resurrection in the concluding chapters of John's Gospel.

Genesis: The Garden of Eden

There are few biblical stories as well known as that of the Garden of Eden. In the Yahwist's story of creation (Genesis 2–3), Eden is a garden planted by YHWH as a delightful and sustaining place for the newly-created Adam (literally, "earth-creature") to live. The river that rises in the garden flows out into four branches, which water and bring life to the whole earth. In the garden are all sorts of trees pleasant to the sight and good for food, but outstanding among them are the "tree of life" and the "tree of the knowledge of good and evil." Amidst all this flourishing abundance, the solitary human being does not begin to find fulfillment until God creates woman. With this development, relational mutuality, generativity, and kinship ties are planted at the center of human joy. Still, the human beings have yet to face their defining moment. YHWH had told Adam, "You may freely eat of every tree of the garden; but of the tree of the knowledge of good and evil you shall not eat, for in the day that you eat of it you shall die." Now a serpent appears and seduces the human beings with the testimony that what will happen if they eat the fruit of the tree of the knowledge of good and evil will be good rather than bad. They eat of it, and suddenly find themselves with an urge to hide from the companionable presence of God. God, noting that the human beings have "become like one of us, knowing good and evil," fears that next they will steal fruit from the tree of life, "and live forever." To prevent this, the human beings are sent out of the garden, and a formidable guard of "the cherubim, and a sword flaming and turning" is placed at the gate to prevent their return.

This story is full of archetypal imagery that has appeared in myriad forms in a variety of world cultures. It has also generated a vast and diverse tradition of retellings, offshoots, and interpretations. Within Western cultures, even those who know almost nothing about the Bible or biblical religions can recount some version of the garden story. Why does such a story of origins in a paradisaical garden, followed by its loss, speak so powerfully to the human psyche and spirit? In the present context I interpret the garden as an image of a deep "belonging in place" for which the human spirit longs. The garden, as represented in this story that is so foundational for Western culture, is wholly a gift of God, and at the same time it is a place that human beings lovingly co-create by tilling it and caring for it. Thus the garden is an earth-place that embraces us and reveals itself to us, even as we embrace it and reveal ourselves to it through our care-taking. We love the place, and the place loves us.

Yet, the human condition as we find it today is such that very few experience such a tender relationship with an earth-place. Whether it be due to the wearing labor demanded of those who get their food directly from the earth, or the ever-renewed waves of human violence through wars and plundering, or the alienation of life in a technologized world, for most human beings "the garden" is a dream rather than a reality. This is exactly the power of the myth of the garden: it enshrines the conviction that we are created for life in the garden, and that despite our present situation of not being there anymore, there is hope that one day a way will be found to have again that experience of earthy rootedness, mutuality, and abundance.

The Song of Songs: The Garden of the Woman

Even within the Bible, other variations of the myth of the garden play upon the same images and themes—yet often with very different effect. One of the most significant of these variations is the garden of the Song of Songs.[3] Since this is not as familiar as the Genesis story, I quote the most relevant verses in full:

> 4:12   A garden locked is my sister, my bride,
>   A garden locked, a fountain sealed.

[3] Sources used include Jill M. Munro, "Spikenard and Saffron: The Imagery of the Song of Songs," *Journal for the Study of the Old Testament,* Supplement 203 (Sheffield, Eng.: Sheffield Academic Press, 1995); Dianne Bergant, *Song of Songs: The Love Poetry of Scripture* (Hyde Park, N.Y.: New City, 1998).

4:13   Your channel is an orchard of pomegranates,
     with all choicest fruits . . . .
4:15   A garden fountain, a well of living water
     and flowing streams from Lebanon.
4:16   Awake, O north wind, and come, O south wind!
     blow upon my garden
     that its fragrance may be wafted abroad.
    Let my beloved come to his garden
     and eat its choicest fruits.
5:1   I come to my garden, my sister, my bride;
     I gather my myrrh with my spice,
    I eat my honeycomb with my honey,
     I drink my wine with my milk.

Whereas in Genesis the woman appeared in the garden as a some-what secondary figure, in the Song the garden *is* a woman; and this woman is of the utmost erotic fascination to the man who loves her. Interestingly, this garden too is closed and locked; but whereas there was a finality to the barred gate of Eden, the connotation this time is that the woman has the power to open the garden gate from within to whom she chooses.[4] Within her, as in Eden, is found the "well of living water" that brings abundance of life wherever it flows. Her beloved effusively extols her wonders—and she responds by inviting him in to eat and drink his fill of all good pleasures.

Thus, at the core of this garden story we can find both a similarity and a difference of plot in comparison to the Genesis story. In both stories the garden is a physical place where human beings find the utmost fulfillment, yet entrance to it is beyond the reach of human powers to attain. The Genesis story embeds a hope for the original peace of the garden in our imaginations, yet ends with an image of divine banishment, which suggests that the human condition is such that we will never find that peace. The Song of Songs, in a sense, picks up where Genesis leaves off and proposes a solution: the garden can be found in the most intimate and fulfilling experience of human love.[5] The erotic pursuit will demand everything of the lover, and even after all his efforts

---

[4] Bergant's commentary does an excellent job of developing this feminist perspective on the woman of the Song.

[5] In this section I am following most contemporary commentaries in interpreting the *Song* as a celebration of human sexual love. In a subsequent section the long tradition of interpreting it as an expression of the intimate mystical relationship between God and human beings will be considered.

he must still await the graceful invitation of his beloved before he can enter. Nevertheless, the mood and take-home message of this garden story could hardly be more different from that of Genesis. The center of gravity has been shifted from a story of alienation from God to one of hope in human intimacy; from God's barring of the gate to the garden to the woman's free choice to open it.

It is important to note that the Song does not offer a naive promise of paradise to be found in some simple sense in sex or in human love. After the lover enters the woman's garden, their enjoyment is not permanent; for in the verses that follow those quoted above, lover and beloved are once again separated and engaged in an urgent search for one another. Indeed, the whole Song is an interwoven dance of presence and absence, desire and celebration, joyous fulfillment and renewal of painful longing. While the overall mood is clearly dominated by the erotic pleasure of the pursuit and its moments of fulfillment, the countertheme of loss, absence, confusion, and unfulfillment keeps on reappearing. This garden is within reach, but a maddening elusiveness remains. True love is not simple, even though it is at least humanly possible. Thus, the Song's vision of human fulfillment is not only gloriously lyrical, but realistic as well. Its potential as a resource for a contemporary spirituality is only beginning to be tapped.

The Gospel of John: The Garden of Passion and Resurrection

In the Gospel of John, a garden allusion appears at the beginning and the end of the passion narrative, and also in one of the resurrection stories. John 18:1 places the locale of Jesus' betrayal and arrest in "a place where there was a garden."[6] The Synoptics, not mentioning a garden, precede the betrayal with a scene of agonizing decision, but John simply presents Jesus as stepping boldly forward to ask whom the soldiers are seeking and then, three times, asserting "I am he." As Peter takes up his sword to resolve the situation by violent action, Jesus rebukes him and instead hands himself over in exchange for the freedom of all his companions. In this garden, then, we can hear an echo of the

---

[6] Matthew and Mark name the place of betrayal as Gethsemane. It is by conflation of their account and that of John that the traditional "garden of Gethsemane" image has arisen.

We should note here also that many contemporary commentators are reluctant to affirm an intentional relationship between Hebrew Bible garden imagery and John's references to gardens in 18:1 and 19:41. For summary see Francis J. Moloney, *The Gospel of John*, Sacra Pagina vol. 4 (Collegeville: The Liturgical Press, 1998) 510–11.

theme of the Garden of Eden as a place where one representing all humanity faces a moment of choice that will radically affect all. But this time, in the words of one commentary, the garden is "almost an anti-garden—a surrealistic inversion of expected qualities of a garden";[7] for rather than harmony, abundance, joyful companionship, and flowing fullness of life, in the garden of the Passion Jesus encounters misunderstanding, violence, intimate betrayal, and destruction of life.

John 19:41-42 concludes the passion story with the statement that "there was a garden in the place where he was crucified, and in the garden there was a new tomb . . . [so] they laid Jesus there." A tomb is intrinsically a place of utmost desolation, yet, on the other hand, at this point in the passion narrative Jesus is once again portrayed as surrounded by a loving community that reverences him with a royal burial.[8] It is essential to recall that in John's theology Jesus' passion has itself been his "lifting up" and glorification. At its climax, as he hung dead on the cross, a soldier's cold spear piercing his side had released the long-withheld waters of life (19:31). Thus, even before the resurrection the garden of the tomb already presents an initial resolution of the broken human condition; it embraces within itself not only the utmost grief, horror, and loss but also the gift of abundantly flowing harmony, peace, and fullness of life that Jesus' offering of self has brought to birth.

A final garden allusion in John occurs at 20:15, when Mary Magdalene mistakes the risen Jesus for the gardener. Some commentators see in this a hint that Jesus is, indeed, the rightful caretaker of the harmonious garden the creator-God intended.[9] In any case, what is most important is that in John the garden imagery has been rewoven within a fresh fabric that picks up on Hebrew Bible themes such as the significance of a specific physical place; the reality of radical human disharmony held in tension with the dream of perfect harmony; and the power of the beloved human body as a source of life. In the next section, we will explore how these classic biblical themes have again been rewoven afresh as new generations of searching human beings claimed the garden archetype as an image of their hope.

---

[7] Leland Ryken et al., *Dictionary of Biblical Imagery* (Downers Grove, Ill.: InterVarsity Press, 1998) 317.

[8] Moloney, *Gospel of John,* 510–11.

[9] Edwyn C. Hoskyns, *The Fourth Gospel* (London: Faber & Faber, 1942) 542; see also Nicolas Wyatt, "'Supposing Him to Be the Gardener' (John 20:15). A Study of the Paradise Motif in John," *Zeitschrift für die Neutestamentliche Wissenschaft* 81 (1990) 21–38.

THE PARADISAICAL GARDEN IN WESTERN CULTURE

In his *History of Paradise,* Jean Delumeau notes that Augustine thought there were three ways of interpreting the story of the Edenic paradise: first, as a corporeal place; second, as a spiritual or mystical reality; or third, as both.[10] Augustine preferred the third option, and was frequently cited by later authors who followed the same approach. I will now briefly follow the track of each of the first two approaches (corporeal and mystical), as well as explore how the garden myth has been taken up into United States cultural ideology and spirituality.

The Garden of Paradise as a Geographical Location

The idea that "Paradise" is an actual place still existing somewhere on earth predates Christianity, and did not really die as a viable mythology until at least the seventeenth century. In the Jewish *Book of Jubilees* (167–140 B.C.E.?), Noah is depicted as dividing the earth by lot among his three sons, giving Shem the portion that includes the Garden of Eden. The Christian Ephraem the Syrian (d. 373) taught that paradise is both a mountain and "like a ring of light" that embraces the whole earth, sea, and sky. During this early period a specific understanding of "sacred geography" came to be commonly accepted in the Western world. Delumeau summarizes: "The earthly paradise is now beyond human reach, either because it sits on an inaccessible height or because it is located beyond an impassable ocean. But it is not therefore unconnected with our earth. It supplies the earth with water (some authors regarding it as the source of the ocean, others as the real though mysterious origins of the great rivers that make life possible in our inhabited world)."[11]

During the medieval era and on into the early modern period, another version of the dream of an earthly paradise had a profound impact. The legend of the idyllic Christian Kingdom of Prester John, far away in the East, appeared in the twelfth century and lasted until the seventeenth.[12] Originally it was said to be located in Ethiopia, later in India. There were also many other reports of "Happy Isles" located in various regions and given various names, for example, St. Brendan's Isle, Brasil, Ophir, etc. Many of the voyages of global exploration during

[10] Jean Delumeau, *History of Paradise: The Garden of Eden in Myth and Tradition* (New York: Continuum, 1995) 18.

[11] Ibid., 44.

[12] Ibid., 76.

the fifteenth and sixteenth centuries were fueled by the desire and ex-
pectation of finding this physical paradise. With the discovery of the
New World of the Americas, some believed that it had finally been located. In Amerigo Vespucci's letter of 1502, for example, he described
the newly discovered Brazil in entirely paradisaical terms. In fact, the
name Brazil seems to come from a Dutch term, *hy bressail* or *"O Brazil,"*
meaning "happy isle."[13] Christopher Columbus, in his report of his
third voyage, affirmed his belief in the character of the earthly paradise—and his conviction that he was getting closer to it, even though
he thought he would not be able to attain it.[14] This whole complex of
ideas also got connected to a story about seven Moorish bishops sail-
ing off to found seven cities, for which the Spanish conquistadors
searched in vain in the interior of the (future) United States between
1530 and 1540 (finding the Grand Canyon instead).

Monastic and Mystical Gardens

The explicitly mystical understanding of the interior location of
paradise has been mainly developed within commentaries on the Song
of Songs. Until recently, most spiritual commentaries on the Song of
Songs employed allegorical or typological hermeneutics that assumed
that although the Song uses intense and explicit imagery of human
erotic love, it really intends to refer to the love between God and the
People of God (and/or the individual human soul). At the same time, the
monks who wrote most of these commentaries lived in a physical setting that was designed to exemplify the paradise myth in architecture.

Monasteries typically had interior gardens in which to cultivate
medicinal plants. It became traditional to ring these gardens with ar-
cades called "cloisters," which then became places for meditation. This
was soon associated with the traditional translation of Song of Songs
4:12: "My sister, my spouse, is a garden enclosed, a fountain sealed
up." This tradition also drew on the image from Genesis 3:24 of paradise being "sealed" after the expulsion of Adam and Eve. Whenever
possible, there was a well in the center of the cloister—an image of the
waters welling up out of paradise to nourish the earth. In the spirituality of the monks, the abundance and harmony of Eden would only be
found on earth by retreating into the *hortus conclusus* (enclosed garden) of the cloister. As Bernard of Clairvaux (1090–1153) put it: "Truly,

---

[13] Ibid., 104.
[14] Ibid., 54.

the cloister is a paradise, a realm protected by the rampart of a discipline that contains a rich abundance of priceless treasures."[15]

Most of the mystical commentaries on the Song of Songs actually dealt only with its first two chapters, never getting as far as chapter 4 where the explicit garden imagery becomes prominent. Bernard of Clairvaux, for example, dwelt much more on the related imagery of the vineyard. The wine of divine union is prepared in the vineyard, stored in the interior wine cellar, and tasted in the ecstatic tryst with the divine lover.[16] The myth that Bernard offers his followers, then, is one of radical interior transformation that literally closes its eyes (in ecstasy) to the physical earth or what happens there.

Four hundred years later, the Carmelite Teresa of Avila still drew deeply upon this tradition of the monastery as a "paradise." She explicitly called her first monastic foundation "a paradise of delight for [God]" and "a heaven, if one can be had on this earth."[17] The imagery of paradise, gardens, wells, flowing streams, and fountains is very central in Teresa's imaginative world.[18] One of her most well-known and powerful images is the allegory in which she compares development in the life of interior prayer to four ways of watering a garden.[19] First one must carry water with great labor; then one learns to turn a water wheel; next water flows abundantly through irrigation ditches; finally rain soaks the garden with no human labor required. At the hinge of modernity, Teresa continued the monastic tradition of understanding the abundant life of the garden as primarily an interior reality that is also imaged and supported by life within an enclosed contemplative community.

The American Dream: The "Garden of the World"

The initial European discovery and exploration of the Americas took place around the same time that Teresa was describing her interior garden. Meanwhile, the hope for an actual earthly paradise was having its final fling, and initial descriptions of the New World sometimes painted it in the colors of the biblical Eden. By the seventeenth century, when

[15] Bernard of Clairvaux, quoted in Delumeau, 121–22.

[16] See, for example, Bernard's Sermon 23 in *Sermons on the Song of Songs*.

[17] Teresa of Avila, *Book of Her Life* 35:12; *The Way of Perfection* 13:7.

[18] Cf. Joseph Chorpenning, "The Monastery, Paradise, and the Castle: Literary Images and Spiritual Development in St. Teresa of Avila," *Bulletin of Hispanic Studies* 62:3 (July 1985) 245–57. For another study of water imagery in Teresa, see Mary Frohlich, "Teresa: Story Theologian and Transformer of Culture," *Review for Religious* 61:1 (January–February 2002) especially 18–20.

[19] Teresa of Avila, *Life*, chs. 11–22.

North American settlement began in earnest, few any longer subscribed to the idea that this land was the actual biblical paradise. But this was also the era when cultivated gardens—an attempt to create one's own personal "paradise"—became much more prominent in European culture. Meanwhile, on the imaginative level, authors were creating fantastic epics in which rural landscapes of delight were places where young lovers never wearied of each other.[20] It was as if the loss of hope in a God-given geographical Eden opened the floodgates on efforts to humanly construct one, either on earth or in the imagination.

These trends took on a unique character as people of European heritage faced the vast wilderness spaces of the New World. According to Leo Marx, by the time of the American Revolution "the cardinal image of American aspirations was a rural landscape, a well-ordered green garden magnified to continental size." Thomas Jefferson and many others among the founding group had the dream that if only each family had its own piece of land to cultivate and make into a garden, this new country could live permanently in a kind of peace and harmony unknown in the Old World. Marx also notes the claim of Henry Nash Smith that well into the twentieth century, "the imagination of Americans was dominated by the idea of transforming the wild heartland into such a new 'Garden of the World.'"[21]

The problem with this ideal, however, was that it included no stopping point. Once there were no more geographical frontiers to conquer, "progress" had to focus on increasing technological mastery of the environment.[22] At some point the core idea of each family prospering happily in its own rural garden tilted over into a nation where farmers are a tiny and embattled minority while most people have only the most tenuous connection to any land at all. As Phil Hefner and Robert Benne put it, the deeply ingrained American ethos of constantly thrusting forward to conquer new frontiers tends to leave behind "our sustaining contexts, our structures of belonging from which we emerged and were given identity . . . the 'place' from which we came."[23]

---

[20] Delumeau, *History of Paradise*, ch. 6.

[21] Leo Marx, *The Machine in the Garden: Technology and the Pastoral Ideal in America* (New York: Oxford University Press, 1964) 141. Marx is referring to Henry Nash Smith's *Virgin Land: The American West as Symbol and Myth* (Cambridge: Harvard University Press, 1950).

[22] Marx, *Machine in the Garden*, 226.

[23] Robert Benne and Philip Hefner, *Defining America: A Christian Critique of the American Dream* (Philadelphia: Fortress, 1974) 44.

RE-APPROPRIATING THE MYTH FOR TODAY

Once again, we find ourselves face to face with the young couple who exemplify the spiritual challenges of life in postmodern America. What can the "myth of the garden" mean for those who have lost touch with sustaining contexts, a sense of belonging, a "place" to come from? Here I will suggest three ways in which young postmodern Americans might be assisted in finding spiritual sustenance through the myth of the garden.

The Shapeshifting Power of Symbols

Reflection on the myth of the garden offers an invitation to jaded postmoderns to rediscover the vitality and dynamism of classical images and stories. Just the awareness that the myth has so many facets could be enlightening for our young couple. This brief survey has shown how the myth of the garden can be a story about belonging and alienation in relation to a beloved place; about the thrill and glory of erotic intimacy; about finding life in the midst of death; about risk-taking journey in search of a paradisaical land; about the heights and depths of mystical ecstasy; about the American dream of wild frontiers made garden-like; and this is only a sampling of the possibilities! For people who feel overwhelmed by the constant pressure of multiple stimuli, pluralistic frames of reference, and general flux, it can be quite grounding to discover how such an archetypal story has the dynamic capacity continually to root itself in new contexts, hybridize with whatever is there, flower into fresh forms, and, in the midst of all this, create energizing links among diverse agendas. This itself may be an image of hope for those whose idea of biblical religion has been that it is peddling an outmoded and static view of the world.

Rewriting the Inner Story of Desolation and Hope

At a deeper level the myth has the potential to invite its hearers into a new exploration of their own inner landscape of fear, grief, love, and hope. The overload of surface imagery from television, radio, Internet, and a fast-paced lifestyle can easily numb the capacity for sensitive awareness and discernment in relation to one's most authentic inner movements. Especially when taken together, the variations on the myth of the garden can offer both affirmation and critique of many of the dreams, fears, and fantasies that stir in the human soul.

One of the most obvious dimensions of this is the desperate need of today's young people for a spiritual vision that affirms their sexual openness while pointing them toward a discerning search for the fullest human and divine potential of sexuality. The ancient mystical traditions envisioned the garden as an interior place of bliss with God, having little obvious relation to ordinary life. Today we can recognize it also as the deep feeling capacity of our bodies, rooted (literally) in the earth and in loving exchange with one another—including, although not limited to, sexual love. A deep reading of the Song of Songs, beginning with its celebratory affirmation of erotic experience and then using the garden theme to explore its relations with the whole network of biblical and mystical themes, could be a way of opening up such a vision.[24]

Today as always, an essential part of the minister's challenge will be to assist people in avoiding the pitfall of identifying their personal "paradise" in too narrow a fashion. The ideal sexual partner; the perfect living space; experiences of mystical ecstasy; freedom to pursue one's wildest dreams; none of these, finally, will be "enough" apart from a deeply rooted yet resilient sense of at-homeness in oneself, one's physical earth-place, one's human relationships, and one's God. This, ultimately, is what the myth of the garden is all about. Yet the risk of identifying this too facilely with personal comfort is always with us. The "dangerous memory" of the Johannine garden of the tomb reminds us that we will not know paradise until our hearts are able to embrace death and sorrow on the way to fullness of life. Although likely to engender resistance in a "Don't worry, be happy" culture, this insight ultimately frees the soul by enabling a vision of wholeness that embraces the real traumas each one has endured.

Spirituality of Place

Finally, and perhaps with greatest urgency, the myth of the garden can be an entree to development of a personal "spirituality of place." Our young couple, like so many in the United States today, rarely feel much relationship to the geographical places in which they move about. As Tom Driver once put it, their focus is not on being in any particular place but rather on getting to "where certain people happen to live, to where they gather, to certain equipment (people plus equipment

---

[24] Dianne Bergant's commentary on the Song of Songs, noted above, offers an outstanding resource for such a catechetical process. Professor Keith Egan of St. Mary's College, Notre Dame, Indiana, however, tells me that in his experience it is easier said than done. Many college students today are so steeped in open sexuality that they can hardly relate to the tender and urgent pathos of the Song of Songs.

equal the office) or where certain events are scheduled."[25] Yet discovery of a connection to the soil, living beings, history, and needs of a specific earth-place is a deeply grounding experience for human beings. It is also an urgent need of the earth itself, whose ecological crisis threatens every member of the earth community. The myth of the garden offers a way of reflecting on the intimate relation of humans as "earth-creatures" to the soil from which we are constructed, as well as the abundant springs of life that can flow from a relationship of mutuality with a beloved earth-place.

For human beings who have established such a relationship, their "garden" (whether as small as a back yard, or as large as the wilderness) can be a cared-for and homey space of beauty and nourishment, full of special delights, a place to dwell with family and to share with others. Rather than being a retreat from the disturbing challenges of postmodernity, this is a necessary foundation for facing them. The *Powers of Ten* video reminds us that today we do not live only on our little patch of grass, but in a teeming city, a pluralistic nation, a threatened planet, a vast universe. Indeed, Philip Hefner has proposed that the present challenge of planetary evolution requires, above all, the creation of cultures that broaden "trans-kin altruism"—that is, that can sanction the search for that homey garden on the broadest scale, and honor others' gardens as much as one's own.[26] Yet without a foundation in a real relationship with the soil and flora and fauna of some specific place or places, it is unlikely that many will find the energy and vision needed for such a project.

CONCLUSION

Perhaps here it is helpful to recall Augustine's reflection that some have interpreted paradise as a physical place; others as an interior mystical place; but the best interpretation is "both." In medieval times, people envisioned the garden of Eden concretely as a geographical place given by God, but now far away and practically unattainable. Today we need to be equally concrete in envisioning it as the whole earth, given to us as gift yet having to be cared for and constructed in collaboration with God, nature, and the many cultures of humanity. The awareness that the garden does, indeed, have to be physically constructed may be

---

[25] Quoted in Benne and Hefner, *Defining America*, 46.
[26] Hefner, *The Human Factor*.

a particular gift that can emerge from the American experience of "subduing the wilderness." Yet it will not be enough only to focus outward; for here and around the world, the search for the inner mystical paradise is still very much alive. Our search as ministers, then, must be for a vision that can adequately convey the "both." The myth of the garden may be a surprisingly apt resource in our efforts to assist our people in discovering the mutual flow between a deeply rooted inner at-homeness in the divine, and the hard communal work of constructing "the Garden of the World."

10

Nature's Parables and the Preaching of the Gospel

MARY CATHERINE HILKERT, O.P.

One of Jesus' most characteristic ways of drawing his hearers into the mystery of God was through his preaching of parables. His preaching strategy was similar whether he was speaking of prodigal parents and wayward children, employers and workers, a woman in search of lost resources, or one who was ethnically despised caring for an injured enemy who had been ignored by religious leaders. In each case Jesus first caught his hearers' attention by describing a familiar life situation. Then his parables took an unexpected turn, which shocked his hearers when, for example, he told of a betrayed parent who rushes out to celebrate the return of a child who has squandered the family's resources before any word of apology or conversion on the part of the son. The Reign of God as depicted in the parables of Jesus includes lavish and undeserved abundance and unconditional forgiveness, repeated reminders that "God's ways are not our ways." The final step in each of the parables leaves the hearers—then and now—with an invitation and a challenge: Do we want to participate in the Reign of God and embrace the conversion that will be required if we are to live by the logic of grace?[1]

[1] For one summary of this interpretation of the parables, see Edward Schillebeeckx, *Jesus: An Experiment in Christology*, trans. Hubert Hoskins (New York: Seabury, 1979) 154–72.

But Jesus did not always draw on human experience to announce the Reign of God in his parables. Often he turned to what the mediaevals termed "the Book of Nature" for his starting point. He used images of seeds randomly sown, dying fig trees, untended vineyards, and wandering sheep as entry points for speaking about God's providence, fidelity, justice, and compassion. As with the other parables drawn from the dynamics of human experience and human relationships, the basic dynamic in those parables remained the same. A familiar pattern (e.g., a vital crop requires careful planning and tending), followed by the disturbing logic of the Reign of God (God's vineyard flourishes in spite of human inattention), leaves the hearers with a decision: Will we throw in our lot with this kind of extravagant and unpredictable God?

Jesus also turned to the Book of Nature to summon his disciples to be attentive to the "signs of the times" in other forms of preaching in the Gospels.

> When you see a cloud rising in the west, you say immediately that rain is coming—and so it does. When the wind blows from the south, you say it is going to be hot—and so it is. You hypocrites! If you can interpret the portents of earth and sky, why can you not interpret the present time? (Luke 12: 54-56).

Dianne Bergant has been one of the prophetic voices in our day who has called us to attend more closely to the Book of Nature and to see the signs of the time in the present moment in evolutionary history. Our growing ecological awareness presents a challenge to conversion not unlike the shift of horizons the parables demand.[2] Both the Book of Nature and the Bible call us to recognize that the Reign of God—the story of salvation—includes not only human history but also all of creation. Bergant's work as a biblical scholar and theologian reminds us that the present ecological crisis presents us with an opportunity to discover aspects of our biblical heritage that previously were overlooked. That rediscovery provides rich resources for contemporary preachers since as Bergant has noted, the preacher's task is to reinterpret earlier (biblical)

---

[2] See, e.g., Dianne Bergant, "Restoration as Re-creation in Hosea 2," in *The Ecological Challenge: Ethical, Liturgical and Spiritual Responses,* Richard N. Fragomeni and John T. Pawlikowski, eds. (Collegeville: The Liturgical Press, 1994) 3–15; idem, "The greening of the tradition: the wisdom tradition and creation," *Theology Digest* 47/2 (Summer 2000) 124–34; and idem, *The Earth Is the Lord's: The Bible, Ecology, and Worship* (Collegeville: The Liturgical Press, 1998).

material within a new context which requires "new expressions of fundamental faith as well as the articulation of new insight."[3]

In order to probe the implications of Bergant's insight for Christian preachers, this essay begins by exploring the new dimensions of the Word of God which come to the fore when preachers attend more closely to the Book of Nature. Then we turn to the question of how ecological spirituality extends the Christian call to hear the cries of the poor to include the groaning of all of creation, a point that Bergant has also stressed in her writing. While the Book of Nature provides an invaluable source for Christian preachers, the final section of the chapter recalls that if preachers are to announce the Gospel, the story of Jesus as reflected in the Scriptures and the liturgy will serve as the lens for the preacher's discovery of the logic of grace in nature's parables.[4]

THE BOOK OF NATURE AS REVELATION OF THE WORD OF GOD

Ecological awareness and concern about massive destruction of the environment have been two of the "signs of our times" that have encouraged Christian believers in prayer and preaching to take a clue from the pattern of the psalms which intertwine God's redemptive activity with God's creative power:

> Your mercy, Lord, spans the sky;
> Your faithfulness soars among the clouds.
> Your integrity towers like a mountain,
> Your justice runs deeper than the sea.
> Lord, you embrace all life: how we prize your tender mercy (Psalm 36).

Preaching on creation as sacrament can move the community to the awe, praise and thanksgiving that spontaneously arises from a deeper awareness of the majesty and glory of God. Here we might learn to see "God's footprints everywhere in creation" (Augustine) from contemplative writers and poets like Annie Dillard and Wendell Berry. No naive naturalist, Dillard's *Pilgrim at Tinker's Creek* includes startling examples of the violence within nature. Nevertheless, she recognizes and

---

[3] Dianne Bergant, "Liturgy and Scripture: Creating a New World," in *Liturgy and Social Justice*, Edward M. Grosz, ed. (Collegeville: The Liturgical Press, 1989) 17. Bergant notes that, in her view, this is precisely the role of biblical preaching.

[4] For further development, see Mary Catherine Hilkert, "Preaching from the Book of Nature," *Worship* 76/4 (July 2002) 290–313.

proclaims what the psalmist named "God's glory" and what the Christian sees as the activity of the Spirit throughout creation. Naming the grace within the ambiguity of creation, Dillard proclaims:

> If the landscape reveals one certainty, it is that the extravagant gesture is the very stuff of creation. After the one extravagant gesture of creation in the first place, the universe has come to deal exclusively in extravagances, flinging intricacies and colossi down aeons of emptiness, heaping profusions on profligacies with ever-fresh vigor. The whole show has been on fire from the word go.[5]

The "original blessing" of creation itself is the major theme of Genesis 1 and the creation psalms, a leit motif which continues throughout the Pentateuch with its focus on the land, and with the proclamation of jubilee as a season of rest and re-creation for the fields of the earth as well as for persons from all social classes. The grace of creation appears in the images of salvation as "new creation" in Second Isaiah, in the wisdom literature, and in the very image of Wisdom as God's darling child at play from the first moment of creation. The Scriptures repeatedly invite us to listen to a preaching we have often responded to with deaf ears: "the heavens declare the glory of God." This proclamation of the heavens needs to be announced more frequently from our pulpits as well.

The human community is invited to join our thanks and praise with the larger gathering of all that shares life and existence, to join our song not only with the communion of saints and angels, but with all of creation. With Job we are reminded of the chorus at song long before human voices joined in: "Where were you when I laid the foundation of the earth? Who laid its cornerstone when the morning stars sang together and all the heavenly beings shouted for joy?" (Job 38:4, 6-7). This is no new-age twentieth-century spirituality, but rather the biblical and liturgical heritage which holds the power to form the Christian assembly into a community that lives more fully in right relation. The Liturgy of the Hours and the seasons of the church year follow and celebrate the rhythms of the cosmos (in the Northern hemisphere). Both Christian and Jewish liturgical feasts can trace their origins to harvest festivals. The climactic celebration of the Christian liturgical year—the Easter Vigil—is replete with celebrations of creation as sacrament: the cosmic images of the *Exsultet,* the blessing of the fire and the paschal candle, the proclamation of the creation text from Genesis, the blessing of the

---

[5] Annie Dillard, *Pilgrim at Tinker's Creek* (New York: HarperCollins, 1974) 11.

baptismal waters and the immersion of the newly baptized, the lavish anointing with oil. The celebration culminates in the consecration of simple resources as both fruits of the earth that have been shaped by the work of human hands and a fragile community of faith are transformed by the Spirit to become the body of Christ. The very celebration of the vigil during the night when darkness slowly gave way to light was perceived as a figure of resurrection in the early Church.

Neither a pantheistic divinizing of nature itself, nor a romantic reading of the story of the universe that fails to deal with the ambiguity and violence within nature "red in tooth and claw," a Christian reading of the Book of Nature is rooted in a theology of creation and incarnation. While preserving the distinction between Creator and creature, Christians proclaim that the divine has become one not only with humankind, but with the material world itself, in the person of Jesus, the firstborn of all creation, whose life, death, and resurrection serve as pledge that from the beginning, all of creation was "destined in love."

This incarnational/sacramental view of the universe can also be traced back to early Christian and mediaeval authors who envisioned all of creation as somehow imaging God. Augustine pointed to the vestiges of God to be found throughout creation and remarked that God had chosen the most commonly needed things of the earth for Christian sacrament. Bonaventure proclaimed that the universe is like a book, reflecting its maker, the Trinity. Drawing out the implications of Bonaventure's claim for the contemporary context, the Australian ecological theologian Denis Edwards has remarked: "Every creature is a revelatory word written in the great book of creation. Every species, every ecosystem, the whole biosphere, every grain of sand and every galaxy, is a self-expression of the eternal Art of divine Wisdom."[6] Nor was that mediaeval insight limited to the descendents of Francis, troubadour of nature and patron of ecology. Thomas Aquinas concurred that all of creation, by its very participation in being, was a manifestation of God. Further, he pointed to the diversity within creation as a created sign of divine goodness which could never be adequately represented by one single creature. A century earlier, the abbess and mystic Hildegard of Bingen, described the presence of the Spirit of God in all of creation more poetically:

> The Spirit of God
> is a life that bestows life

---

[6] Denis Edwards, *Jesus the Wisdom of God: An Ecological Theology* (Maryknoll, N.Y.: Orbis Books, 1995) 109–10.

root of the world-tree
and wind in its branches.
She is glistening life
alluring all praise
all-awakening
all-resurrecting.[7]

Like human experience, however, the revelation of nature, does not always provide a clear manifestation of the mystery of the God whom
Jesus revealed to have a preferential love for the poor and suffering. If
preachers of the Gospel are to turn to creation for a word of grace, they
need to listen closely to the cries of the poor and the groaning of all of
creation and to discover the relationship between the two.

THE CRIES OF THE POOR
AND THE GROANING OF THE EARTH

There is no denying that the world we inhabit and proclaim as
graced is at the same time a history of radical suffering. That history
encompasses not only human suffering but also a world of ecological
devastation. The social teaching of the Church has challenged believers to move beyond an individual or interpersonal ethic and to reread
the Gospel with an awareness of how social and ecclesial structures
can either bind or liberate human persons. Now our era of global ecological threat and destruction calls us to extend that mandate and think
again about what it means to be created as relational beings who are
finite and interdependent, about who is our neighbor, about what our
relationship with the earth is meant to be, about who or what constitute the "little ones" who are the focus of God's compassion.

Of particular concern is the dichotomy that exists in many people's
minds between social justice and ecological justice. All too often proposed solutions to environmental destruction do not take account of
the inequitable burden those solutions place on the backs of the poor
including loss of the jobs and resources that have been their source of
survival in the present structures of the global economy. Pope John
Paul II's New Year's message for the celebration of the World Day of
Peace on January 1, 1990, identified the ecological crisis as a moral
problem and called Christians to realize "their responsibility within

---

[7] In *St. Hildegard of Bingen: Symphonia: A critical edition of the Symphonia armonie
celestium revelationum,* Barbara Newman, trans. and ed. (Ithaca, N.Y.: Cornell University
Press, 1988) 140–41.

creation and their duty towards nature and the Creator are an essential part of their faith."[8]

In the early 1990s the United States' Roman Catholic Bishops underscored the links between social justice and ecological justice and specifically called for "greater attention to the extent and urgency of the environmental crisis in preaching."[9] Insisting that the "web of life is one," they remarked that "generations yet unborn will bear the cost for our failure to act today" and further, "in most countries today, including our own it is the poor and powerless who most directly bear the burden of current environmental carelessness."[10]

The interconnection of the web of injustice is illustrated further by the racism one can detect as also involved in decisions to locate toxic waste dumps or incinerators with toxic fumes in poor neighborhoods where the population most frequently is comprised of African-American, Hispanic/Latino/a, or a variety of immigrant populations. In preaching from the perspective of creation as sacrament, the preacher is called to show how the destruction of the environment causes even further human suffering especially among the poor and most vulnerable.

But even that is not enough. Rather, if all of creation reflects the glory of God, the cries of other threatened forms of life signal the violation of God's beloved creation in itself and not solely because it threatens human life. Both preachers and communities of believers are called to reflect on how social and ecological injustice are interrelated effects of a common logic which focuses on individualism, competition, acquisition, and seeking one's own comfort and advantage without regard for the common good. In both social injustice and ecological devastation, a common logic is at work: the logic of the marketplace. As the Brazilian liberation theologian Leonardo Boff observed in his book *Cry of the Earth, Cry of the Poor*: "The very same logic of the prevailing system of accumulation and social organization that leads to the exploitation of workers also leads to the pillaging of whole nations and ultimately to the plundering of nature."[11]

[8] John Paul II, "The Ecological Crisis: A Common Responsibility," Message for the Celebration of World Day of Peace, January 1, 1990, in *"And God Saw That It Was Good": Catholic Theology and the Environment,* Drew Christiansen and Walter Grazer, eds. (Washington, D.C.: United States Catholic Conference, 1996) 215–22 at 222.

[9] United States Bishops' Conference, "Renewing the Earth," *Origins* 21:27 (December 12, 1991) 432.

[10] Ibid., 426.

[11] Leonardo Boff, *Cry of the Earth, Cry of the Poor* (Maryknoll, N.Y.: Orbis Books, 1997) 110–11.

Prophetic preachers will call the community to see those connections, to hear the lament of the earth, to recognize our role in causing the devastation of nature. The call to repentance and conversion requires a new way of seeing the relationship between human creatures and the rest of creation that involves both kinship and responsibility. The move to an identity and ethic of stewardship, while an important advance over earlier approaches to what it meant for human creatures to be commissioned to exercise "dominion" over the rest of creation (Gen 1:28), does not capture the radical and mutual interdependence that exists between human beings and all other creatures and life forms. In an age of unprecedented ecological disasters, the prophecy of Isaiah takes on new meaning:

> The earth mourns and fades, the world languishes and withers;
> both heaven and earth languish.
> The earth is polluted because of its inhabitants,
> who have transgressed laws,
> violated statutes, broken the ancient covenant.
> Therefore a curse devours the earth, and its inhabitants pay for their guilt
> (Isa 24:4-6).

PREACHING A NEW LOGIC OF GRACE

Whether addressing social justice, ecological justice, or both, the preacher's role is to proclaim an alternative vision of reality that operates by a different logic—the logic of grace. In the gospel vision of reality, all are included at the table, relationships operate by the logic of forgiveness, the last come first, masters wash feet, the little ones are cared for, resources are shared and in the sharing, multiplied, and the earth reveals the glory of God. In the economy of God, the poor proclaim good news and the land shares in a "year of favor."

How does the preacher invite the assembly into that kind of parable world? An obvious place to start is by using stories drawn not only from human experience, but also from nature—as Jesus did in his stories of a mother hen, a fig tree, or a mustard seed. If we but attend, nature and science provide incredibly rich images of unexplained growth, of miraculous recovery, of care for the young and vulnerable, of possibilities of life on planets or in places human beings had declared uninhabitable. If the Jesus of John's Gospel could use images of grains of wheat dying to bring forth new life and if Paul could turn not only to human sleep but to seeds and plants for analogies for the resurrection, why do

Christian preachers so rarely turn to the mysteries of the universe to describe what the Reign of God is like?

But not all the parables of nature disclose mercy and compassion, or provoke awe. Just as Jesus included the complexity and even sinfulness of human experience in his parables drawn from human experience—unjust judges, workers who murdered the vineyard owner's son, the unscrupulous manager—so the preacher who pays attention to the world around her will detect parables drawn from creation's stories of violence, devastation, and destruction. Preachers who contemplate nature as faithfully as Dillard did, will stumble upon some version of the giant water bug's attack on a small green frog. As Dillard described the violence she witnessed:

> [The heavy-bodied brown bug's] grasping forelegs are mighty and hooked inward. It seizes a victim with these legs, hugs it tight, and paralyzes it with enzymes injected during a vicious bite. That one bite is the only bite it ever takes. Through the puncture shoot the poisons that dissolve the victim's muscles and bones and organs—all but the skin—and through it the giant water bug sucks out the victim's body, reduced to a juice. This event is quite common in warm fresh water.[12]

If a story of grace is to be found there, it will require preachers to wrestle for a blessing as they reflect on the events within nature that at the very least *seem* to contrast with the God of life whom Jesus enfleshed and proclaimed. Christians see parables of grace at work in the universe, and wrestle with parables of dis-grace, in light of another story that provides the key for interpreting the creation story—the story of Jesus proclaimed in the Scriptures and celebrated in Christian worship. The story of the cosmos remains ambiguous; there are many versions of its origins and future. Preachers of the Gospel turn to the Book of Nature searching for both an echo and new versions of another story whose final word is always life. The life, death, and resurrection of Jesus provide the lenses for a Christian preaching of the story of the universe as a story of grace.

On the one hand, the Book of Nature provides an ecological lens that reinterprets and expands the Scriptures and liturgical celebrations with a new appreciation of the cosmic dimensions of sacramentality. The Book of Nature sharpens contemplative sensitivity to the incomprehensibility of the mysteries of the universe as revelatory of the incomprehensible mystery of God. Nature's text expands our horizon of

---

[12] Dillard, *Pilgrim at Tinker's Creek*, 8.

praise and thanks beyond God's mercy and compassion to God's chosen people to God's love for all of creation. Hearkening to nature's word widens the scope of our ethical awareness and moral challenges to include ecological justice and to require new forms of asceticism if life in all of its diversity and uniqueness is to continue, let alone flourish. An often-overlooked source of revelation, the Book of Nature offers a vast new source of parables for Christian preachers and gatherings of faith.

On the other hand, just as the Christian Scriptures need to be reinterpreted in light of contemporary scientific and ecological awareness, the "story of the universe" also needs to be interpreted. What are we to make of the waste, and violence in creation? Are entropy, randomness, and chance the final words about the universe and its purpose? What are the ethical lessons we are to learn from nature's ways? Survival of the fittest? Natural selection? Evolutionary history to date does not promise future fulfillment for all of earth's creatures, especially for the vulnerable and the weak.

But Christians read the Book of Nature in light of another story that we believe is God's "dream of the earth" proclaimed and enfleshed by Jesus. As herald of the good news of the Reign of God Jesus announced that despite all the evidence to the contrary—in nature as well as in human history—flourishing is the final destiny of all life. In his life, ministry, death, and resurrection, Jesus offered an alternative reading not only of human life, but of all of life as interrelated, and of a power of love at work throughout creation. His reading of reality included the potential of compassion to change stories that seem inevitable. Resources that appeared limited and inadequate—fish and bread—were available in abundance when they were shared rather than hoarded. The power at the heart of the universe that he named as "Abba" was a God of life, not destruction. Again and again, he revealed an energy of love—the power of the Spirit—at work in the world, healing human bodies and spirits, casting out demons, and calming chaos in both human life and nature. The Reign of God he preached, the divine will he enfleshed, disclosed a God of mercy and compassion, who sides with the poor and oppressed, who stands in solidarity with those who suffer, who grieves dying, loss, and violence, who promises life. Jesus' radical solidarity with the vulnerable and the outcast even unto death limits the shape of a Christian reading of the creation story and focuses the Christian telling of the parables of nature.

In the context of a global crisis we have hardly imagined, Christian communities and preachers are commissioned as disciples of the one who came to bring life, "life in all its fullness" (John 10:10). The parable of Jesus' life and preaching culminates in the parable of his death and resurrection. Nature's own parables reveal, but also challenge, the paschal claim that life, rather than death, is the final destiny for all of God's beloved creation. In the story of the universe, fragile life is often snuffed out by predators and survival of the fittest is at the expense of the vulnerable. But even in nature it would appear that kinship and an instinct to protect "the little ones" at times triumph over self-interest and even survival. An article in *National Geographic* several years ago recounted one such example discovered by a forest ranger in Yellowstone National Park. An anonymous preacher connected the story with Psalm 91 in the following passage circulated on the Internet:

> After a forest fire in Yellowstone National Park, forest rangers began their trek up a mountain to assess the inferno's damage. One ranger found a bird literally petrified in ashes, perched statuesquely on the ground at the base of a tree. Somewhat sickened by the eerie sight, he knocked over the bird with a stick. When he gently struck it, three tiny chicks scurried from under their dead mother's wings. The loving mother, keenly aware of impending disaster, had carried her offspring to the base of the tree and gathered them under her wings, instinctively knowing that the toxic smoke would rise. She could have flown to safety but refused to abandon her babies. When the blaze had arrived and the heat scorched her small body, the mother remained steadfast. Because she had been willing to die, those under the cover of her wings would now live. "he [sic] will cover you with his feathers, and under his wings you will find refuge" (Ps 91:4).[13]

While the mother bird's actions may derive from evolutionary instincts, with human consciousness and freedom come the capacity and responsibility of creatures of the universe to consciously "choose life" or "choose death."

The hope of Christians—and of all Christian preaching—turns on the claim that Jesus' death was not the end, that the Spirit of love restored the dead Jesus to a transformed life. That same love which moves the sun and the stars promises transformed life for all forms of life that have been defeated or destroyed. If the testimony of the first Christians is true, if God has indeed raised Jesus from the death, then there can be a future for the rest of creation as well. Rooted in the conviction that the incarnation involves the union of divinity not only with

---

[13] "Wings," author and source unknown.

humanity, but with material creation, Karl Rahner has pointed out that Easter is the feast of the future of the Earth (and the entire cosmos):

> [Christ] rose not to show that he was leaving the tomb of the earth once and for all, but in order to demonstrate that precisely that tomb of the dead—the body and the earth—has finally changed into the glorious, immeasurable house of the living God and of the God-filled soul of the Son. He did not go forth from the dwelling place of earth by rising from the dead. For he still possesses, of course, definitively and transfigured, his body, which is a piece of the earth, a piece which still belongs to it as a part of its reality and destiny. . . . Already from the heart of the world into which he descended in death, the new forces of a transfigured earth are at work.[14]

Using an agricultural image, Paul reminded the Corinthian community that the resurrection of Jesus provides the first fruits of the final harvest of creation. Grounded in the faith conviction that the life, death, and resurrection of Jesus confirm the victory of life over the forces of death, Christians proclaim a hope that is not only human hope but ecological hope as well. Resurrection faith announces that God has taken on the powers of death and the gates of hell, and broken their power. In Jesus' confrontation with the forces of death, God pledged solidarity not only with vulnerable human life, but also with all creatures who have been considered expendable.

Neither the history of humankind nor the history of evolution provide clear evidence that hope for the future is warranted if that hope is to include the vanquished. But resurrection faith is grounded finally neither in human commitments nor in confidence in the evidence of natural processes, but in the power of the Creator God who gave birth to the cosmos in its beginnings, who draws life out of death and redeems the lost, and whose Spirit is the source of all life moving through the universe. For Christians, God is revealed definitely in the final parable of the resurrection of Jesus which establishes a future not only for human persons, but for all of creation. In that historical and cosmic parable, the totally unexpected has happened with the overthrow of the power of sin and death. The one who brought the world into existence with a word of love has spoken a final promise of life through the power of a love that even death cannot quench. The challenge to preach—and live by—that logic of grace is ours.

---

[14] Karl Rahner, "Easter: A Faith that Loves the Earth," in *The Great Church Year* (New York: Crossroad, 1993) 195.

11

Preaching Morality Without Moralizing

ANDREW L. NELSON

INTRODUCTION

The Sunday homily, the focus of this brief study, is intended to interpret the assigned readings of the day in ways that are meaningful for the lives of our people, and illuminative regarding their corresponding moral responsibilities. With increasing frequency and sharpening focus, official post-conciliar documents promote a particular approach to the moral content of Sunday homilies. This approach features two closely interrelated themes: the good news and morality, corresponding to the biblical themes of Gospel and Law.

In this essay I will first provide brief examples that incorporate both themes, explain why preaching is deficient whenever these are not interrelated, and offer reflections of biblical scholars on the nature of their interrelationships. I will attempt to identify the approach implied in the documents and suggest a specific method of preaching morality that builds upon it. To this end, I will propose a simple, two-step procedure as one way to address the intent of this approach, and specifically, a manner in which moral issues may be properly incorporated in a homily. It will feature two guiding concerns or two "moments" of the homily:

1. The homily should proclaim the Good News, i.e., that aspect of the wonderful works of our God on our behalf, the history of salvation culminating in the Christ-event, and continuing through the work of the Holy Spirit, that is found in the readings—what we may entitle here the Divine initiative; and

2. The homily should identify and urge those moral responsibilities, whether ongoing conversion, or specific moral perspectives, attitudes or behavior, that follow from this divine initiative—here named our graced response. (Ours will always be differentiated from the original responses, good or ill, in the texts, but nonetheless attentive to them.)

PREACHING MORALITY

Two brief but representative quotations leave no doubt in our minds that the homilist must indeed preach morality. Second Timothy is described as "an earnest pastoral letter from a veteran missionary to a younger colleague, who is responsible for a group of churches and for preserving them from destructive influences from without and from dissidents from within."[1] The citation below, clearly a moral exhortation following upon a presentation in doctrine, is typical of the Pauline corpus. The author, perhaps Paul himself, has every reason to counsel Timothy to preach morality, modeled upon his own ministry of the Word. He bases this counsel upon his prior specific proclamation of faith in the God of salvation history.

> I, [Paul], charge you, [Timothy], in the presence of God and Jesus Christ, who will judge the living and the dead, and in view of his appearing and his kingdom. Proclaim the word; be persistent, whether the time is favorable or unfavorable; convince, rebuke, and encourage, with the utmost patience in teaching. For the time is coming when people will not put up with sound teaching, . . . Always be vigilant, endure suffering, do the work of an evangelist, carry out your ministry fully (2 Tim 4:1-5).

Timothy has been called to be "a gospel-herald with a Spirit-empowered challenge, accosting the [Christian community]."[2] To rebuke and encourage, to challenge and accost, and with persistence in the face of stubborn opposition, are duties with profound moral meaning. Moreover, these duties are predicated upon the preacher's convictions about God's sustaining presence, God's kingdom.

---

[1] Warren Quanbeck, "Introduction to II Timothy," *The New Oxford Annotated Bible* (New York: Oxford Press, 1994) 306.

[2] Joseph Fitzmyer, *To Advance the Gospel* (New York: Crossroad, 1981) 155.

St. Gregory the Great (d. 604), pope and doctor of the Church, is often hailed as the original fashioner of moral theology. He insists:

> Pastors who lack foresight hesitate to say openly what is right because they fear losing people's favor. . . . The Lord reproaches them through the prophet: "They are dumb dogs that cannot bark." . . . They are afraid to reproach people for their faults and thereby lull the evildoer with an empty promise of safety. . . . They keep silent and fail to point out the sinner's wrongdoing. Anyone ordained a priest undertakes the task of preaching, so that with a loud cry he may go ahead of the terrible judge who follows. If, then, a priest does not know how to preach, what kind of cry can such a dumb herald utter?[3]

Gregory has little patience with the preacher who fails to encourage his people in sound doctrine and is too timid to raise controversial issues or to identify peoples' pressing moral responsibilities. For him, whoever does not preach morality fails to protect and prepare his people for their ultimate accountability before God. Such a dereliction of duty renders one a "a dumb dog that cannot bark," a "dumb herald."

BUT NOT MORALIZING

With the *Catechism* we may define a homily as "an exhortation to accept this Word, not as a human word, but as what it truly is, the Word of God, which is at work in you believers (1 Thess 2:13) and to put it into practice."[4] Yet, sometimes it is simply an annoying harangue! In such a sacred responsibility there lurks the danger of exaggerating or isolating the moral component of one's preaching, of descending into a diatribe, of dissociating the moral mandate from the overarching saving action of God.

Moses may well have been guilty of such anger in word and deed, revealing both his lack of trust in God and his excessive anger with the people of God. Moses was supposed to assemble the complaining, thirsty people before the rock so that they could again witness God's power, love, and deliverance. He was to trust that God would continue to intervene on behalf of his people. He also was to speak to the rock. Instead he uttered words of serious doubt and administered a harsh

[3] St. Gregory, *Pastoral Guide*, n. 4, Gregory the Great. Trans. from *Liturgy of the Hours*, 4:343–44.

[4] Interdicasterial Commission, *Catechism of the Catholic Church*, 2nd ed. (Washington, D.C.: United States Catholic Conference, 2002) n. 1349.

accusation against the people, "Hear now, you rebels: shall we bring forth water for you out of the rock?" Then, as if acting out his anger, he struck the rock twice with his rod. In words of reproach, Moses not only departed from God's directive, but also "twisted what was intended as an opportunity for the community to witness a creative word-event of God's power and love into a word-event of doubt and hostility. God's gift was accompanied by resentful, spiteful words, instead of power-filled, gift-bearing words."[5]

Moses here displayed a tragic failure in faith while severely misrepresenting this act of God's intervention. His words led the people to turn an event of God's graceful deliverance into an action of reproach. God then said to him: "Because you did not believe in me, to sanctify me in the eyes of the people Israel, therefore you shall not bring this assembly into the land which I have given them" (Num 20:12).

The penalty imposed was indeed severe—never to lead his people into, or even set foot upon, the land of promise and destined to die in the wilderness. However, we should not underestimate his blunder. Without the firm conviction that God's love and care always preceded and accompanied him, his anger and doubt dominated him. These turned his words into an unwarranted, insulting, and contentious reprimand. His words and actions exemplify "moralizing" at its ugliest.

MORALIZING EXPLAINED

According to a 1994 document from the Pontifical Biblical Commission (PBC),[6] the homilist should explain "the central contribution of the texts, that which is most enlightening for faith and most stimulating for the progress of the Christian life." The Word of God is to become "more and more the spiritual nourishment if the people of God, and of love. . . ." The document goes on to offer this challenge: "Want of preparation in this area leads to the temptation to avoid plumbing the depths of the biblical readings and to be content simply to *moralize* or to speak of contemporary issues in a way that fails to shed on them the light of God's word" (IV:C3, emphasis added).

Used here, "moralizing" means lecturing about some aspect of morality that is one's particular "crusade" or interest. This is not

[5] Rita Burns, *Exodus, Leviticus, Numbers,* Old Testament Message 3 (Wilmington, Del.: Michael Glazier, 1983) 256.

[6] Pontifical Biblical Commission, *The Interpretation of the Bible in the Church* (September 21, 1993) *Origins* 23:29 (January 6, 1994).

preaching a morality that is the fruit of serious reflection and study of the encompassing biblical message. Without the illumination, critique and motivation that only the Word of God can provide, such a moral message is at best incomplete, and at worst, an abuse of sacred space and time. "Moralizing" suggests that one is "preaching" from an exclusively moral/ethical perspective, i.e., from within a narrow, legalistic context, emphasizing heavily people's duty to obey. The doctrine and practice of religion are subtly being reduced to morality. In this understanding, primacy belongs to the law, and ultimately to God as the divine lawgiver. As a consequence, people are led to regard their lives as one dimensional—circumscribed by the obligation to obey the mammoth collection of laws. Persons in this mindset must labor continually under the fear of divine disapproval and rejection, deprived of the comforting security and acceptance of a loving and forgiving God.

The cognate "moralistic" often conveys the imposition of an inflexible, rigorist piety—similar to the stance of 1 Timothy's opponents who forbade marriage and demanded abstinence from many foods. Moralistic teaching, then, would reflect a predominantly negative assessment of human nature, viewed as severely compromised by concupiscence, requiring a constant penitential rigor, and best motivated by the fear of eternal damnation. Here Jesus would be portrayed as a stern and inscrutable redeemer. Such moralizing, significantly remote from the message of God's love, is always a tendency among teachers and preachers. Rather than life giving, hopeful, and a source of inspiration, this type of message is often harsh and accusatory, even intimidating if not actually condemnatory.

A RELATED LESSON FROM HISTORY

In the seventeenth century the subject matter of the new academic study of moral theology was organized into a distinctive text, the moral manual. At first this text consisted of little more than an extended commentary on the second part of the widely accepted *Summa Theologiae* of Thomas Aquinas. Gradually, as its subject matter increasingly focused on the specific concerns arising from priests' ministry in the sacrament of penance, the manual began to incorporate and classify the varying arguments and opinions of recognized moralists. The orientation grew increasingly act-centered, even sin-centered, burdened with attempts to evaluate specific sins according to solidly probable moral opinions. This intensely practical moral enterprise, originally an integral part, now became distanced from the first and third parts of Thomas' *Summa*. The comprehensive context, the doctrinal vision, and system of Thomas and his commentators were left behind. The burgeoning tradition of dogmatic or systematic theology, anchored in Sacred Scripture and the Fathers, ceased providing the moorings for moral theology. Instead the new discipline was evolving into a philosophy of natural law, casuistry, and ecclesiastical positivism.

The law was presented here largely as an independent reality, or as an ordinance focused on the moral order decreed at creation. Moral teaching was being severed from salvation history, and particularly, from the Christ-event. The usual recourse to Sacred Scripture was simply to provide "proof-texting." The discipline became oriented toward law, sin, and confession, and closely aligned with the discipline of canon law. It would retain this orientation into the middle of the twentieth century.

The Roman Catholic professors of the Tübingen School of the early nineteenth century constructed a renewed moral teaching upon a retrieved biblical orientation, influenced, no doubt, by their Protestant colleagues on the faculty. In fact, the "kingdom of God" became a prominent reference point in the work of Johann Baptist von Hirscher. Recourse to the biblical themes was later to influence Bernard Häring, the prominent Redemptorist moralist and *peritus* at the Second Vatican Council, who had studied in Tübingen.

What the council came to decree concerning moral theology is especially illuminating for the approach to moral preaching advocated after the council. Moral theology "should be renewed through a more lively contact with the mystery of Christ and the history of salvation. Its presentation, drawing more fully on the teaching of holy scripture, should highlight the lofty vocation of the Christian faithful and their obligation to bring forth fruit in love for the life of the world" (Decree on the Training of Priests, n. 16).

A renewal was demanded. The longstanding preoccupation with law and sin, which had underwritten the moralizing that often dominated Sunday sermons, was finally recognized as both barren and uninspiring. Instead, the primary concerns were to be one's calling from God, as presented in Scripture, and one's moral responsibilities that flow from this divine vocation.

This brief conciliar directive would initiate a major revision in the discipline. The "rupture" between moral teaching and the foundational doctrines of salvation history, the incarnation and redemption, and the abiding presence of the Holy Spirit was finally under repair.

Only now is the discipline of moral theology reclaiming a biblical foundation, and an integral relationship with the prominent dogmatic themes rooted there. Only now is this revised moral theology postured to assist the homilist in considering moral issues from within the larger context of God's invitation and loving purpose.

GOOD NEWS PRECEDES OBLIGATIONS

This responsibility of preaching morality, then, is no simple task. It requires something other than an emotional eruption, a scolding or a bold confrontation. The Sunday homily should express Christian joy. Hence,

> we homilists must do out part to insure that the news we proclaim is *good news*. Homilies which leave people guilty and dispirited are by nature not Sunday homilies. We proclaim the good news of the resurrection and of our inclusion into Christ through baptism. We proclaim the good news of the world charged with the grandeur of God.[7]

Returning to the PBC document, we read:

> Preachers should certainly avoid insisting in a *one-sided* way on the *obligations* incumbent upon believers. The biblical message must preserve its principal characteristic of being *the good news of salvation freely offered by God*. Preaching will perform a task more useful and more conformed to the Bible if it helps the faithful above all to "know the *gift* of God" (John 4:10) as it has been revealed in Scripture; they will then understand in a *positive light* the *obligations* that flow from it (IV.C3, emphases added).

Here "one-sided," negative moralizing is contrasted with the grandeur of proclaiming the marvelous news of our God's generous gift. In fact, because the homilist first stresses this divine gift, the obligations of the faithful become discernible, in their positive light, as responsive to the salvation that God is offering.

GOSPEL AND LAW, INTERRELATED

What is the "Gospel" that we are to present—the defining context for whatever moral teaching? Does it stand in solitary splendor, or does it somehow imply or even anticipate our moral response? It is "the

---

[7] Stephen DeLeers, "The Homily as Sunday's Word," *New Theology Review* 14 (November 2001) 78.

message about God's new mode of salvific activity on behalf of human beings made present in Jesus Christ, his Son." It contains six salient characteristics, according to a study by Joseph Fitzmyer. The Gospel is:

1. Revelatory—a new manifestation of God's saving activity, now through the Lordship of Jesus Christ;
2. Dynamic—"a salvific force unleashed by God in human history through the person, ministry, passion, death, and resurrection of Jesus, bringing with it effects that human beings can appropriate by faith in him," and closely associated with the Holy Spirit;
3. Related to the pre-existing, primitive kerygma, not only as to content and the actual proclaiming, but especially as it re-presents Jesus to people, as one who continues to confront us "with God's new mode of saving activity to be appropriated by faith working itself out through love";
4. Promissory—a concrete realization of God's promises of old;
5. Universal—God's saving action in Christ is intended for everyone;
6. Normative—"stand[ing] critically over Christian conduct, church officials, ecclesiastical teaching, and even the written Scriptures themselves, . . . tolerating no rival, . . . it accosts and challenges [people] to conform to its proclamation." It provides inspiration and guidance for the Christian community. It also liberates believers from any imposition of humanly constructed legalisms.[8]

This rich portrayal of the Gospel, including its challenging norms, must be understood by the homilist. These two need not be regarded as starkly contradictory, or in dialectical tension. They are clearly not mutually exclusive categories. In fact, a credible treatment of "law" in the Old Testament also articulates this intimate relationship. The Torah itself was considered a gift of a loving God.

> The Law is closely related to Israel's self-understanding as a covenant community under God. *Divine grace and mercy* are the presuppositions of law in the OT; and the grace and love of God displayed in the NT events issue in the legal obligations of the New Covenant . . . Torah means guidance, direction, pointing the way for the faithful Israelite and for the community. Not merely the laws of Pentateuch provide guidance; the entire *story of God's dealings with Israel* points the way . . . In the OT then, law is understood to rest on *the initiative of God*, who has redeemed a particular people, entered into covenant with them, and provided the basic stipulations of this covenant relationship with covenant law. The *love and*

---

[8] Derived from Fitzmyer, 152–58.

grace of God constitute the *setting* for the giving of the law. *Responsive love*, gratitude, and faith provide the motivation for obedience to the law.[9]

The presentation of the law in the New Testament maintains this relationship with the divine initiative, here explained as the kingdom of God at hand: "By setting the call to repentance in the context of the givenness and immediacy of the kingdom, Jesus freed it from mere moralism and utterly radicalized it."[10]

The call to repentance, whereby one is to embrace the law of love, flows from one's acknowledgement of *God's prior loving act* of inaugurating the kingdom. Jesus presents this law, not as some arbitrary imposition of a demanding deity, but as rooted in salvation history.

The New Testament documents "find in the law both the passing shadow of *the gospel* to come and that which is completed or fulfilled 'in Christ.' They affirm that the law, insofar as it is the expression of the holy will of God, remains valid, radicalized, and at the same time relativized, by the absolute claim of love."[11]

The law, then, as presented in both testaments, is to be grasped within the larger, enveloping context that may well be identified as salvation history, or as the Reign of God's absolute love, the central element of Jesus' preaching, or simply, as the Gospel.

THE TWO MOMENTS IN THE HOMILY

A brief survey of pertinent official documents may assist us in appreciating the PBC's directive regarding the context for preaching morality. An explicit relationship between the moral directives for believers and the overarching saving action of God began to emerge with increasing clarity beginning with conciliar teaching and extending into the present. Once they had established the preeminence of Sacred Scripture in the Liturgy, and correspondingly in the sermon, now called homily, the conciliar documents regularly mentioned two central moments, variously described as (1) the proclamation of God's wonderful works in the history of salvation, and (2) the mystery of Christ ever made present and active in us; or as (1) the Mysteries of the faith, and

---

[9] W. J. Harrelson, "Law in the Old Testament," *The Interpreter's Dictionary of the Bible* (Nashville: Abingdon Press, 1962) 3:88–89, emphases added.

[10] W. D. Davies, "Law in the New Testament," *Interpreter's Dictionary of the Bible* (Nashville: Abingdon Press, 1962) 3:97.

[11] Davies, 102, emphasis added.

(2) the guiding principles of the Christian life (Constitution on the Liturgy, nn. 24 and 52).

These are also expressed in another, similar fashion as (1) the invisible God, from the fullness of his love, addresses men and women as his friends and lives among them in order to invite and receive them into his own company, and (2) the obedience of faith must be our response to God who reveals. By faith one freely commits oneself entirely to God (Dogmatic Constitution on Divine Revelation, nn. 2 and 5).

The 1982 publication of the National Conference of Catholic Bishops *Fulfilled in Your Hearing*[12] states that the liturgical homily draws on the Scriptures to interpret people's lives in such a way that they can recognize the saving presence of God and turn to him in praise and thanksgiving (26). "Such doctrinal instruction and moral exhortation . . . are situated in a *broader context*, namely, in the recognition of God's active presence in the lives of people, and the praise and thanksgiving that this response elicits" (26). Homily preparation groups should share the good news, . . . God's promise, power, and influence, and share the challenge (to conversion) that these words offer us (26–27).

In PBC we find this passage regarding the homily: "It is fitting to explain the central contribution of the texts, that which is most enlightening for faith and [that which is] most stimulating for progress in the Christian life" (IV.C3).

Moral instruction surely is an integral component of the homily, but not as the sole focus, and never without reference to God's saving activity, to which moral living is a response. To preach morality in a constructive manner, then, we must emphasize not one but two moments: the always prior loving divine initiative and the always subsequent human response. We may best understand these in terms of a dialogical or "relational/responsible" model. The following sets may illuminate these moments:

Gospel/Kerygma	*Didache/Catechesis*
What has God done for us?	What are we to do?
God's self revelation/communication in Jesus Christ	human response of faith and active discipleship
Divine gift (always already offered)	acknowledged gratefully and actualized
Divine forgiveness	re-creation or new creation

---

[12] The Bishops' Committee on Priestly Life and Ministry, *Fulfilled in Your Hearing* (Washington, D.C.: United States Catholic Conference, 1982).

| Indicative: God loves us as God's own . . . | Imperative: Let us behave accordingly . . . . |
| Protasis "If God has done this . . . ." | Apodasis "Then we in turn must . . . ." |

If we think of the relationship of the two moments as dialogical, then our moral teaching is clearly perceived as flowing from, and interacting with, the specific good news that we have proclaimed. The directives and demands are obviously grounded in and endorsed by the Gospel. Whether we are calling people to forgiveness, peace and justice, compassion, or thanksgiving, whether to engagement or withdrawal, we must make clear that the attitudes, behavior, perceptions, and identity that we are promoting are to be in some way reliant on the good news that is here proclaimed.

Josef Jungmann, in his commentary on deliberations over the Constitution on the Sacred Liturgy's treatment of the homily, reminds us that the tradition of simply preaching catechetical sermons did not die easily: "Some [council fathers] desired free scope for the consecutive treatment of the catechism segments [which were] customary in some countries or generally for the non-homiletic sermon, a desire which also found expression at the final voting in some of the modi."[13]

While ongoing catechesis would always be endorsed in general, the council fathers determined that the homily has a prior purpose. It was to render the proclaimed word of God intelligible and meaningful. The more traditional type of preaching—it could hardly be called homilizing since it was not biblically based—was to be replaced by one which always begins with the scriptural texts.

AN UNCOMPLICATED METHOD

A method for examining the three Sunday readings and the psalm response in order to preach morality without moralizing is offered here. The following three questions could be posed to these lectionary passages:

1. What has God been doing here? (the divine initiative)
2. What are respondents called to do or become? (the original response)
3. What are we called to do or become? (our graced response)

[13] Josef Jungmann, "Commentary on the Constitution on the Liturgy," in *Commentary on the Documents of Vatican II*, Herbert Vorgrimler, ed. (West Germany: Herder and Herder, 1967) 38.

These questions help to identify the dynamism of the texts, common themes, and the overarching divine-human dialogue. This procedure may open the homilist to new possibilities for upholding the two moments in the texts, and for preaching morality without moralizing.

As the Gospel is preached appropriately, people are enabled to identify or discern in their own experience the authentic call of God, either in their innermost depths, or beneath their conscience's evaluation of their own behavior, attitudes, or mindset. They may recognize God's call in the blessings already received in the form of their Christian faith, instances of personal deliverance or even their own talents, skills, and predispositions. They may also be assisted in hearing God's call in the corrections and challenges that come their way, in the love and forgiveness they must extend and receive from family and friends, and in the joy and peace they must foster and experience. We must help them interpret these as graces and summons to give thanks to God, to follow God's lead, to respond generously.

Each preacher of morality must be a student of the sacred word, and of various procedures of interpretation.[14] Homilists must preach first of all to themselves, lest they be unmasked as frauds. They are to strive for candor, self-understanding, keenly aware of their own journey of faith and trust, their own weaknesses and blunders, their own need for reconciliation. In their hearts they must find thanks for God's blessings, for their part in salvation history and the life of the Church of Christ, for the grace of ministry and the remarkable, undeserved blessing to preach the Word of God and call their sisters and brothers to sanctity.

CONCLUSION

We are not called to be "dumb heralds" as St. Gregory reminds us. Of course we must preach morality, sometimes boldly, always courageously. We must challenge, accost, rebuke, encourage, and inspire. We must identify the evil in our midst, whether it be malice or mediocrity, violence or complacency, obstinacy or erroneous thinking. As Timothy was called, so must we always "be vigilant, endure suffering, do the work of an evangelist, carry out your ministry fully" (2 Tim 4:5). But we

[14] Respected commentaries are of immense assistance. Dianne Bergant's three-volume *Preaching the New Lectionary* is especially attentive to the issues addressed in this paper. See Dianne Bergant, with Richard Fragomeni, *Preaching the New Lectionary*, 3 vols. (Collegeville: The Liturgical Press, 1999–2001).

must do so without rancor, prejudice, or harsh accusation. We must do so within the larger context, first proclaiming our God's constant love and mercy that anticipate our responses, the Divine initiative. Whether the readings present God as consoling or confronting us, God is always acting out of love. Even our most challenging homily must give prominence to this conviction.

Contributors

BARBARA E. BOWE, R.S.C.J., is professor of biblical studies and director of the Biblical Spirituality Certificate Program at Catholic Theological Union She holds the TH.D. from Harvard University. She serves on the editorial boards of *The Catholic Biblical Quarterly, The Bible Today,* and the Filipino journal *Daʾan (The Way).* She has published in *The Bible Today, Missiology, The Journal of Early Christian Studies, Theological Education, U.S. Catholic,* and other journals. Her newest book is *Biblical Foundations of Spirituality* (2003).

MARY C. BOYS, S.N.J.M., is the Skinner and McAlpin Professor of Practical Theology at Union Theological Seminary in New York City. She holds the ED.D. from Columbia University and Union Theological Seminary. She is past president of the Association of Professors and Researchers in Religious Education (APRRE). Her most recent book is *Has God Only One Blessing? Judaism as a Source of Christian Self-Understanding* (2000).

WALTER BRUEGGEMAN is the William Marcellus McPheeters Professor of Old Testament at Columbia Theological Seminary. He holds the TH.D. from Union Theological Seminary and PH.D. from St. Louis University. He is interested in interpretive issues that lie behind efforts at Old Testament theology. This includes the relation of the Old Testament to the Christian canon, the Christian history of doctrine, Jewish-Christian interactions, and the cultural reality of pluralism. A prolific author, he has over sixty books to his credit including *Texts That Linger, Words That Explode: Listening to Prophetic Voices* (2000) and *Deep Memory, Exuberant Hope: Contested Truth in a Post-Christian World* (2000).

AGNES CUNNINGHAM, S.S.C.M. S.T.D., has been a member of the Congregation of the Servants of the Holy Heart of Mary since 1943. She completed her studies in theology at the Facultés Catholiques (L'Institut Catholique) Lyon, France (1963–1967). She was professor of patristic and historical theology at Mundelein Seminary (University of Saint Mary of the Lake), Mundelein, Illinois (1967–1992). She continues to study and translate documents for her congregation in Champaign, Illinois, where she is also superior of an active apostolic community.

Carol J. Dempsey, o.p., ph.d., is associate professor of theology at the University of Portland, Oregon. She is an associate editor of the *Catholic Biblical Quarterly* and was a biblical consultant on the *Columbia River Pastoral*. Her most recent publications include *Hope Amid the Ruins: The Ethics of Israel's Prophets* (2000), *The Prophets: A Liberation Critical Reading* (2000), and *Jeremiah: Preacher of Grace, Poet of Truth* (at press). She lectures widely on topics related to Hebrew Bible, in particular, the Prophets, biblical ethics, and ecology.

Edward Foley, capuchin, is professor of liturgy and music and the founding director of the ecumenical D.MIN. program at Catholic Theological Union. A former president of the North American Academy of Liturgy, he was also a member of the founding executive team of the Catholic Academy of Liturgy. He has published thirteen books and numerous pastoral and scholarly articles.

Mary Frohlich, r.s.c.j., ph.d., teaches spirituality at Catholic Theological Union. Much of her work has been on retrieving the Carmelite classics for today's spiritual needs. More recently she has also been exploring the potential of "spirituality of place" as an antidote to postmodern alienation from the natural world. Recent publications include an essay in *Spiritus* on methodology for the study of spirituality; several scholarly articles on Thérèse of Lisieux; and an anthology of Thérèse's writings for the Orbis Books' "Spiritual Masters" series.

Anthony J. Gittins, c.s.sp., ph.d., is the Bishop F. X. Ford, m.m., professor of missiology at the Catholic Theological Union. A cultural anthropologist, he has written numerous books and articles including *A Presence that Disturbs: A Call to Radical Discipleship* (2002), *Encountering Jesus: How People Come to Faith and Discover Discipleship* (2002), *and Ministry at the Margins: Strategy and Spirituality for Mission* (2002).

Mary Catherine Hilkert, o.p., is professor of theology at the University of Notre Dame and vice president of the Catholic Theological Society of America. She is author of *Speaking with Authority: Catherine of Siena and the Voices of Women Today* (2001), *Naming Grace: Preaching and the Sacramental Imagination* (1997), and coeditor with Robert Schreiter of *The Praxis of the Reign of God: An Introduction to the Theology of Edward Schillebeeckx* (2002). She has published numerous articles on contemporary theology, spirituality, and preaching.

Andrew L. Nelson was ordained in 1957, earned an S.T.L. from the Gregorian University, Rome (1958), and a PH.D. in religious studies from Marquette University (1981). He has taught moral theology at Saint Francis Seminary, Milwaukee, since 1978, and there served as academic dean, vice-rector, and rector. Now retired from administration, he is an assisting priest in a Milwaukee parish while continuing part-time teaching at the Seminary and lecturing on topics in moral theology and preaching.

HERMAN E. SCHAALMAN was ordained a Rabbi at Hebrew Union College-Jewish Institute of Religion in Cincinnati. Among his many honors are honorary doctorates from Hebrew Union College—Jewish Institute of Religion, Catholic Theological Union, and Spertus Institute of Judaic Studies. He has served as president of the Chicago Board of Rabbis, the Central Conference of American Rabbis, and the Jewish Council on Urban Affairs. He was chair of the advisory dommittee of the American Jewish Committee and is a member of the education committee of the National Holocaust Council. He has published theological articles in various journals and coedited *Preaching Biblical Texts.*

ROBERT SCHREITER, C.PP.S., was awarded a D.TH. by the University of Nijmegen. He is the Vatican II Professor of Theology at Catholic Theological Union and conjointly the professor of theology and culture at the University of Nijmegen. Among his many publications are *Constructing Local Theologies* (1985) and *The Ministry of Reconciliation* (1998).

RICHARD J. SKLBA holds the licentiate in Sacred Scripture from the Pontifical Biblical Institute in Rome (1964). He was present for the opening session of Vatican Council II (October 1962) and took the formal oath administered by Pope John XXIII to be faithful to the teachings of the council. Former teacher of Sacred Scripture at Saint Francis Seminary in Milwaukee and past president of the Catholic Biblical Association of America, he was ordained an auxiliary bishop in his home archdiocese of Milwaukee in 1979. Since 1998 he has co-chaired the national Lutheran/Catholic dialogue.

Index